Famous Animals in History
and Popular Culture

Famous Animals in History and Popular Culture

ANN C. PAIETTA *AND*
JEAN KAUPPILA

McFarland & Company, Inc., Publishers
Jefferson, North Carolina

Jean Kauppila would like to say a special thank you
to Beth and John Daynes for helping her
to complete this book for Ann C. Paietta who passed away in 2018.

LIBRARY OF CONGRESS CATALOGUING-IN-PUBLICATION DATA

Names: Paietta, Ann Catherine, 1956– author. | Kauppila, Jean L., 1957– author.
Title: Famous animals in history and popular culture / Ann C. Paietta and Jean Kauppila.
Description: Jefferson, North Carolina : McFarland & Company, Inc., Publishers, 2023 |
Includes bibliographical references and index.
Identifiers: LCCN 2022062109 | ISBN 9781476673813 (print) ∞
ISBN 9781476635538 (ebook)
Subjects: LCSH: Famous animals. | BISAC: NATURE / Animals / General | SOCIAL SCIENCE /
Popular Culture
Classification: LCC QL793 .P35 2023 | DDC 590—dc23/eng/20230106
LC record available at https://lccn.loc.gov/2022062109

BRITISH LIBRARY CATALOGUING DATA ARE AVAILABLE

ISBN (print) 978-1-4766-7381-3
ISBN (ebook) 978-1-4766-3553-8

Front cover images: mother elephant with calf © Gino Santa Maria/Shutterstock;
A three-year-old chimpanzee, named Ham, in the biopack couch for the MR-2 suborbital
test flight. January 31, 1961, (NASA); USCGC *Campbell* (WPG-32) crew and mascot,
Sinbad, in the North Atlantic, 1943 (USCG Historian's Office)

Printed in the United States of America

McFarland & Company, Inc., Publishers
Box 611, Jefferson, North Carolina 28640
www.mcfarlandpub.com

Table of Contents

Introduction

Humans have developed close relationships with many animal species over the past 20,000 years, changing both humans and animals in the process. When dealing with animals, people's best and worst qualities are revealed, and their attitude towards animals can often be a strong indicator of their nature. Members of the animal kingdom have long played a role in human history. In this book, you will discover some of the many creatures whose list of achievements range from astronaut to warrior to faithful companion to entertainer to rescuer to explorer to therapist. These are just a few roles animals have played in human culture and history. If only animals could talk, they could tell their own stories of historical events in wartime and peace. Fortunately, humans have recorded animal contributions as participants and witnesses to human history.

Animals and Wars

Until the introduction of modern machinery, animals played an often-decisive role in warfare. Humans have enlisted animals to help fight wars since prehistoric times, and some of the world's earliest historical sources speak of battles between ancient warlords in horse-drawn chariots.

Elephants, with their massive size and impressive tusks, have been used in warfare since ancient times. Elephants were first utilized in military operations in India, and have been used throughout history since then. Famous generals including Pyrrhus of Epirus, Hannibal, and Alexander the Great used elephants to conquer their opponents. War elephants could carry troops or heavy materials, or charge and intimidate.

Pigeons have been used to carry messages since at least the sixth century BCE, when the Persian King Cyrus is said to have used pigeons to communicate with the distant parts of his empire. Pigeons have continued to carry messages throughout human history. One of the most famous wartime pigeons, named Cher Ami, earned the French military decoration Croix de Guerre for delivering twelve messages even when wounded. Between 1943 and 1949, thirty-two pigeons were awarded the Dickin Medal for gallantry or devotion to duty while serving in a military conflict. Pigeon G.I. Joe (1943–61), serving the U.S. Army Pigeon Service, won the medal for saving over 1,000 lives.

Camels were first tamed as pack animals and as herd animals for milk and meat in North Africa and the Middle East around 3,000 years ago. Camels flourished in the deserts, mountains, and vast areas of wastelands. Camels were first recorded being used in war in 853 BCE, by the Arab King Gindibu when he fielded 1,000 camels in an army against the Assyrians at the Battle of Qarqar, in modern day Syria.

Dogs have played a variety of roles during wartime since antiquity—from sentry duty to delivery tools, messengers, search and rescue, and detecting poisonous gases. Here is one example of a failed dog-related project during World War II, when the German Army invaded the USSR in 1941: The Russians were short of anti-tank mines and weapons, so they devised a plan of strapping explosive belts onto the backs of dogs and training them to go under enemy tanks. As you would expect, the dogs panicked, running back under their own tanks, or the Germans just shot the dogs. In another sad war-related story, about 5,000 dogs went to Vietnam as guard dogs or scouts; not a single one came back. The dogs were deemed excess equipment and left behind—many were euthanized or given to the Vietnamese army, and some were left to fend for themselves. In other wars, dogs were treated a little better. In 1992, President Bill Clinton signed legislation that would ensure all military dogs be treated like true veterans.

Horses have played the greatest role of all the animals in the history of warfare. During World War I, in which ten million soldiers died, eight million military horses died as well. Whether pulling chariots, transporting equipment or carrying people to battle, the horse has seen more action in wars than any other animal. In just one day during the 1916 Battle of Verdun in France, some 7,000 horses were killed. Horses are also more susceptible to the elements, and thousands succumbed to exhaustion, disease and poison gas attacks.

Some creatures are without name but played a role in warfare. Bees were used by ancient Greeks and Romans in war by catapulting beehives over the walls of besieged cities. In World War I, soldiers used European glowworms to brighten the dark trenches by collecting them in jars. In the 1960s, the U.S. Navy first began its militarization studies on dolphins, which evolved into a major program training dolphins to deal with underwater mines. Other animals used during wartime include pigs, rats, bats, sea lions, cattle, mosquitos, scorpions, and mules.

Countless animals have served as scouts, guards, messengers, transportation and beasts of burden during wartime. Many more have filled vital roles as mascots and buddies for the troops. Units in all branches of the U.S. military and other countries have adopted mascots, usually dogs, which have lived and often died alongside their human companions. Many commanding officers said yes or just looked the other way when these animals were "adopted" by the troops, knowing how important animals were to morale. During the Civil War, mascots included dogs, horses, a dignified bald eagle named Old Abe, and other creatures. These animals inspired the troops and served as reminders of their own beloved pets at home. Mascots brought loyalty and enthusiasm to the troops and offset boredom in camp. Unfortunately, many in this diverse array of comforting creatures were also casualties of war.

The People's Dispensary for Sick Animals Dickin Medal was instituted in 1943 in the United Kingdom by Maria Dickin to honor the work of animals in World War II. The award is sometimes referred to as the animals' Victoria Cross. It is awarded for conspicuous gallantry or devotion to duty while serving in military conflict. The Dickin Medal has been awarded 74 times, plus one honorary award made posthumously in 2014 to the "heroic war horse" Warrior on behalf of all the animals that served in World War I.

As the modernization of long-distance weapons and technology developed, the use of animals in war has become less common and in most cases has disappeared.

Animals in Space

Animals have been used in space exploration since 1783, when the Montgolfier brothers sent a sheep, a duck, and a rooster up in a hot air balloon. The U.S. used high-altitude balloon flights from 1947 to 1960, testing radiation exposure, physiological responses, life-support and recovery systems. These early flights carried fruit flies, mice, hamsters, guinea pigs, cats, dogs, frogs, goldfish, and monkeys to heights up to 144,000 feet.

In the United States, many primates were launched into space, between 1947 and 1961, in order to study the biological effects of space travel. Species that were used included rhesus monkeys, squirrel monkeys, pig-tailed macaques, and chimpanzees.

In the 1950s, the Soviet Union launched dogs on various suborbital flights. Stray female dogs were used since they were thought to be capable of handling extreme cold and their waste would be easy to control. A Soviet Union dog became the first living being to orbit the Earth, on *Sputnik* 2 on November 3, 1957.

Animals in Entertainment and Sports

Horse racing, rodeos, circuses, and zoos have played a role in entertaining humans, often at the expense of animals. There is a long, rich history socially and economically when it comes to sports and entertainment with regard to these animal-related activities.

Pets

Keeping pets gives many people companionship, pleasure and great happiness. Many breeds of animals have a long history of being human companions. Pets have been found in nearly every culture and society. Pet-keeping apparently satisfies a deep, universal human need. Next to dogs, horses and cats are the animals most intimately associated with humans. The bond can be extremely strong, and in some cases animals have given their lives for their human companions. Everyone knows that animals are capable of receiving love and affection, and there are countless stories of them being rescued from shelters or dire situations. And in return, animals have demonstrated the capacity of returning the favor by showing bravery and rescuing humans in extraordinary ways. Animals have reciprocated by saving human lives (of both family and strangers) and providing affection and companionship. There are even stories of undomesticated animals saving people from imminent death, even when it would appear that there was no reason to do so.

United States presidents and their families have often had pets while serving in office. Pets have helped presidents get elected or re-elected by adding a sense of warmth and humor to campaigns. Pets have included horses, dogs, cats, birds, raccoons, bears, rabbits, goats, sheep, and a hippo, to name just a few. Theodore Roosevelt, the 26th president, was famous for his family's many pets. Some of the more unusual pets have been gifts from other world leaders. Presidential pets have been eyewitnesses to many historical events behind the closed doors of the White House.

Animals in Culture

Animals have been inspiring artists since the first cave paintings of deer, horses, and bison—and pet cats, dogs, birds, and others have inspired creativity with their beauty, humor, companionship, and cuteness. Artists such as Albrecht Durer, George Stubbs, and Edwin Landseer are known for their portraits of animals. Frida Kahlo kept many pets—multiple monkeys, Xoloitzcuintli ("Mexican Hairless") dogs, parrots, parakeets, macaws, chickens, an eagle named Gertrudis Caca Blanca (Gertrude White Shit), and a fawn named Granizo. Kahlo often painted herself surrounded by her animal companions, which were a calming constant.

Literary history is packed with humans who helped inspire writers' creative urges. But not all muses are humans. This role has also been filled by furry or feathered friends. Edgar Allan Poe, Ernest Hemingway, Elizabeth Barrett Browning, Robert Burns, and Gertrude Stein have all been inspired by animals.

Animals further play a wide variety of roles in literature, film, mythology, and religion.

Therapy Animals

Dogs and cats are most commonly used in pet therapy. However, fish, guinea pigs, pigs, birds and other animals can also serve as aides to people. The type of animal chosen depends on the therapeutic goals of a person. Often the purpose of pet therapy is to help someone recover from or cope with a health problem or mental disorder.

For this book, animals are defined as those with vertebrates, such as mammals, birds, fish, reptiles, and amphibians. Animal representations, like college mascots or Punxsutawney Phil the groundhog, are not covered. International coverage of real animals is arranged alphabetically by the first letter of the first name. Years covered are from ancient times to 2015. This book contains over 500 named animals with biographical entries. Included are a bibliography, more than 80 photos, and a subject index.

The animals in this book are all part of a long tradition of animals that have impacted humans and their development in many ways. Not all the animal stories are particularly heart-warming, but all are worthy of their own recognition. Today, social media fuels many animal tales, but the majority of the animals in this book have stood the test of time and have been celebrated throughout the years. There has always been some fictionalization of true animal stories. Unfortunately, the internet in the present day has led to widely circulated tales of animal courage that are pure fiction. For example, there is the story of Daisy the dog, who saved many people during 9/11, but when the story is compared to the facts, it is pure fiction. Some of the stories may be legends, and the facts have probably been greatly embellished over the years, but the tales serve to remind us of the invaluable service of animals to humans.

Subject Entries

Abdul *see* **Murphy**

Able (Monkey)

    Able was a female rhesus monkey born at the Ralph Mitchell Zoo in Independence, Kansas. Able and Miss Baker, a squirrel monkey, were launched on a successful mission aboard the space rocket *Jupiter* AM-18 on May 28, 1959. They traveled at a maximum altitude of 300 miles and downrange of 2,000 miles at speeds of 10,000 mph. Each monkey was fitted with a specially made suit with sensors to track pulse, body temperature and movement. Each was trained to press a button whenever a red light flashed, so that her coordination was tested. Upon their return home, Able and Miss Baker were treated as celebrities. Both monkeys survived the trip well, but Able died in surgery to remove an infected electrode on June 1, 1959. After her death, she was preserved

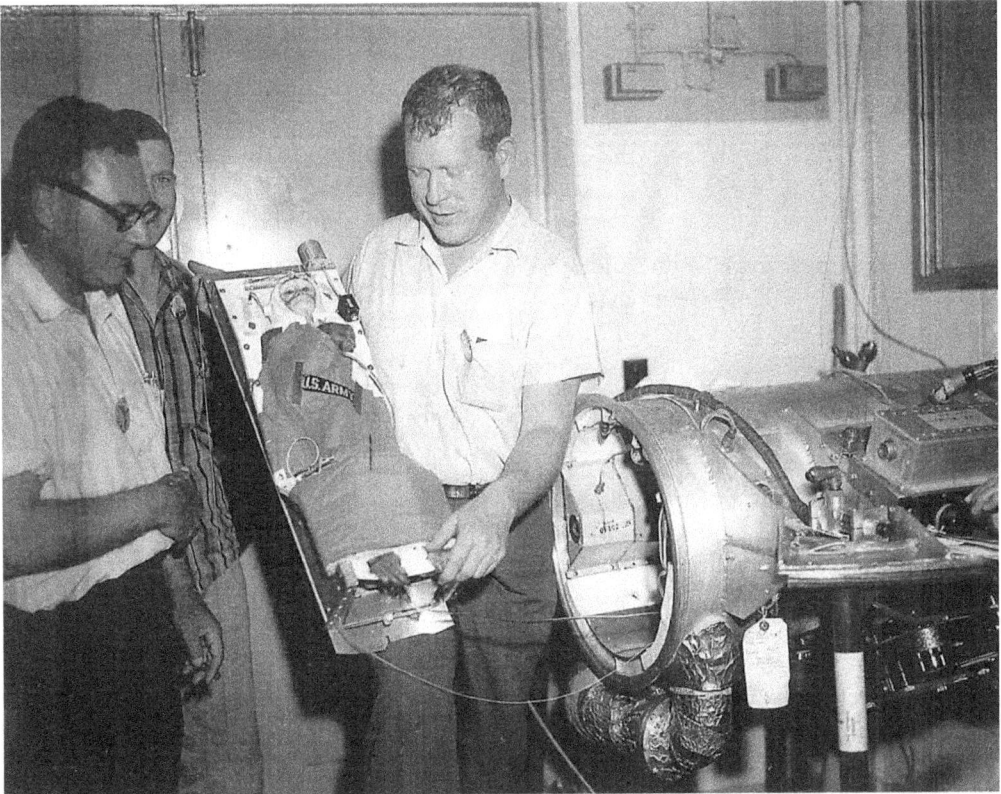

Able preparing for her mission aboard Jupiter AM-18, May 28, 1959 (NASA).

and put on display at the Smithsonian Institution's National Air and Space Museum.

Able Seacat Simon *see* **Simon**

Abul-Abbas (Elephant)

A Jewish trader, Isaac, was sent by Charlemagne, king of the Franks and Holy Roman emperor, to obtain an elephant from Harun al-Rashid of Baghdad. After a four-year mission to the Persian Empire and Africa, Isaac returned to Europe in 801 CE, successfully completing his mission. Seeking friendly relations, Harun al-Rashid sent Abul-Abbas the elephant. Abul-Abbas was housed near the emperor's capital and taken on tours of the empire. There is little dependable history on either Charlemagne or Abul-Abbas. What is known is that the elephant was taken to war with the Danish in 804 CE, but never directly into the fighting. In 810 CE Abul died in his forties of pneumonia, partly due to his regular dips in the Rhine River and the lack of exotic-animal veterinarians.

Adwaita (Tortoise)

Adwaita (meaning "the one and only" in Bengali) was a male Aldabra giant tortoise that lived in the Alipore Zoological Garden of Kolkata, India. Adwaita was reportedly given to Robert Clive (1725–1774) of the East India Company by British seafarers who captured him from Aldabra, an atoll in the Seychelles. Adwaita was transferred to the Alipore Zoological Garden in 1875 or 1876 by the founder of the zoo. Adwaita lived in his enclosure in the zoo until his death in 2006. His age at death was estimated to be as old as 255 years. His life spanned much of modern Indian history. Estimates put his birth date around 1750.

Ahbhu (Dog)

Ahbhu, a Puli or Hungarian sheepdog, was the late author Harlan Ellison's loving companion, which he rescued from the L.A. Animal Services West Los Angeles Center. He was named after a character from the 1941 movie *The Thief of Bagdad*. Ahbhu was the inspiration for the canine character Blood in Ellison's science fiction novella *A Boy and His Dog*. The story follows fifteen-year-old Vic and his telepathic dog, Blood, as the two navigate through a post-apocalyptic United States as survivors of a nuclear World War III between America and the Soviet Union. *A Boy and His Dog* was a winner of the Nebula Award for Best Novella and a Hugo award finalist and adapted into a cult classic film. Ellison said he wrote the story of Ahbhu to demonstrate "the nobility of canine companionship."

Albert I (Monkey)

On June 11, 1948, a V-2 Blossom launched into space carrying Albert I, a rhesus monkey. This unsung animal astronaut hero never received the fanfare due him. He died, and subsequent animals were named Albert and then numbered. Not naming the animals individually made it less personal. Albert II attained an altitude of 83 miles on June 14, 1949, using a second V-2 Blossom rocket, but died when his parachute failed to open. Albert III and Albert IV died during their flights in 1949. In 1951, Albert V also suffered a malfunction with his parachute.

Alex (Parrot)

Alex was an African grey parrot who, from 1977 to 2007, was the subject of an experiment by animal psychologist Irene Pepperberg (University of Arizona, Harvard University, Brandeis University). Alex was purchased from a pet store when he was about one year old. Pepperberg's work shattered the beliefs by most scientists that only a large primate brain could handle problems with language and understanding, and that birds are not as intelligent and are only able to communicate with each other by mimicking and repeating sounds. Pepperberg's work with Alex supported her hypothesis that birds are intelligent creatures that can reason on a basic level and use words. She also wrote that Alex's intelligence was similar to dolphins, great apes, and a five-year-old child, and that he exhibited the emotional level of a two-year-old child. Alex was trained using a "model/rival technique." During his life, Alex achieved many accomplishments: he could identify 50 different objects; recognize quantities up to six; distinguish seven colors and five shapes; understand the concepts of "bigger," "smaller," "same," and "different"; and "speak" a "vocabulary" of over 100 words. Alex died on September 6, 2007, at the age of 31. Some scientists are skeptical about Pepperberg's findings, believing that Alex was getting subtle clues from the questioner, simply imitating his teacher or performing from rote.

Alfred (Gorilla)

During most of his life, Alfred was thought to be a mountain gorilla but was more likely a western lowland gorilla. In 1928, the American Museum of Natural History in New York City and Columbia University found Alfred on an expedition to the Belgian Congo. The expeditioners were told about a pair of gorillas shot by a farmer, and afterwards the baby was suckled by a local woman. The baby gorilla was bought by a Greek merchant; in 1930 he was sold to an Italian who in turn sold him to an animal dealer in Europe, and eventually he made his way to the Bristol Zoo Gardens on September 5, 1930. Alfred was named after Alfred Moseley, a Companion of Honour and benefactor of the zoo. The gorilla became one of the zoo's main attractions. For his first two years at the

zoo, he walked around the grounds wearing woolen jumpers. Many of Alfred's visitors recollect his quirky personality. Among his dislikes were bearded men, double decker buses, airplanes, and Rosie the Elephant walking by his enclosure. One of his habits was forming and throwing snowballs, and two of his more disgusting habits were picking up large lumps of his droppings and hurling them at spectators, and urinating on visitors. Alfred's popularity was spread even further during World War II when U.S. Army troops sent pictures and stories about the gorilla back to their homes. Also, American newspapers continued writing articles featuring Alfred throughout the war. Alfred died on March 10, 1948, from tuberculosis. After his death, taxidermist Rowland Ward mounted Alfred, posing him on all fours, and he was put on display near the Bristol City Museum Cafe. The Anatomy Department at Bristol University received his bones and organs. In 2011 Alfred's status as mascot for the city of Bristol helped to launch the "Wow! Gorillas" project to raise awareness about the extinction crisis facing primates.

Algonquin (Pony)

Algonquin, an eight-year-old Shetland pony, was a gift to President Theodore Roosevelt's nine-year-old son Archie from Secretary of the Interior Ethan Allen Hitchcock. Algonquin was imported from Iceland and remained an endearing White House pet during Roosevelt's presidency, from 1901 to 1909. *The Washington Post* described Algonquin as "iron gray or dun with spots and compactly built with round barrel, small ears, clean pony face, and stocky limbs."

In 1908, the *Washington Evening Star* reported:

There is no home in Washington so full of pets of high and low degree as is the White House, and those pets not only occupy the

Algonquin pictured with President Theodore Roosevelt's son Archie, June 17, 1902 (Frances Benjamin Johnston Collection, Library of Congress).

attention of the children, but the president is himself their good friend, and has a personal interest in every one of them.

Alice (Dog)

Alice was found hungry and abandoned on a wharf on Guam in 1950. Given the rank of O.D. (Ordinary Dog), Alice served during the Korean War in the early 1950s on the destroyer HMCS *Cayuga* for two tours of duty. One of her adventures happened when she fell into the water and was rescued by a sailor who risked his life by being lowered down to pull her out of the water to safety.

All Ball (Cat)

In 1984 All Ball, a gray and white Manx, became the famous gorilla Koko's pet kitten. Koko was the gorilla at the San Francisco Zoo that understood over 1,000 hand signs and more than 2,000 English words. She took care of All Ball like he was her own baby. Tragedy happened in December 1984 when All Ball escaped from a cage at a research facility near Stanford University and was hit by a car and killed.

Andre (Seal)

Andre, a male harbor seal pup, was found at Robinson's Rock in Maine by Harry Goodridge, a 45-year-old tree surgeon who had a side business taking on skin-diving assignments. Harry was motoring out towards Robinson's Rock, a well-known spot for harbor seals to go for sun and rest, searching for a seal pup to be his skin-diving companion. In his book, *A Seal Called Andre*, he wrote about how he came to get Andre: "I saw the small, sleek head fifty feet dead ahead, only its dome and round eyes above water. The seal pup raised his head as if to get a better view. The eyes that met mine showed no alarm. I looked around for the pup's mother. No sign of her. Then a curious and totally unexpected thing happened. Instead of submerging, the pup swam directly towards the boat. I swooped down with my net and swung the little orphan aboard." Thus began a special 25-year relationship. Harry had many other animals over the years, but Andre seemed to be extra special. Andre swam with Harry, went sledding, splashed around in the bathtub and watched *Flipper* on television. Harry trained him to shake hands, hide his eyes with his flipper, and, like a dolphin, leap through a motorcycle tire. In Andre's first year, he spent nearly every day at the house but would sometimes disappear for the whole day, coming back at night. There was a winter when the harbor froze over that Andre disappeared for five months, and whatever he experienced in the wild is unknown. Harry was afraid for Andre's life from ice in the winter and also from the lobstermen and fishermen who often clashed with Andre. Harry began to take Andre to an aquarium for the winter. For six winters, he was at the New England Aquarium in Boston, and he spent four winters at the Mystic Aquarium in Connecticut. Between April and October, Andre would put on a show every day in Rockport, Maine. Andre is buried on Harry's property, and a granite statue carved by Jane Wasey is placed near Rockport's boat landing. In 1994, Paramount Pictures released a feature film, *Andre*.

Andre, a male harbor seal pup, with his life-long companion Harry Goodridge in 1974 (photograph taken by Lew Dietz courtesy Goodridge Family Archives, CC BY-SA 4.0).

Andy (Dog)

Andy, a purebred Doberman pinscher, was nicknamed "Gentleman Jim" by the Marines of M Company because of the proud way he walked and his unwavering dependability. When he became a Marine, the loyalty and obedience he felt for his owner, Theodore Andrea Wiedeman of Norristown, Pennsylvania, was transferred to his two handlers: PFC Robert E. Lansley, of Syracuse, New York, and PFC Jack Mahoney, of Clinton, Connecticut. In 1943 the three of them were sent across the Pacific Ocean to Bougainville Island. The U.S. Marine invasion took the Japanese by surprise; two hundred and fifty Marines followed Andy through the trail from the beaches toward the mountains. The handlers found Andy easy to read, and they did not hesitate to hit the ground the instant Andy halted and went rigid seconds ahead of an explosion. Because of Andy, a Japanese sniper, an enemy patrol, and a camouflaged battery were destroyed along the trail. "Two hundred and fifty Marines went into that jungle. Thanks to Andy of the K-9 Corps, two hundred and fifty Marines came out."

Annie (Dog)

Annie the Railroad Dog makes the Roadside Pet Cemetery list of heroic American dogs that are kind, hardworking mutts that seem to appear out of nowhere to win the hearts of people. In 1934, Annie, a collie mix, was adopted by the railroad men at the Mason Street train station in Fort Collins, Colorado. Annie became somewhat of a local celebrity as she greeted arriving passengers. There is a book about her life, *Annie, the Railroad Dog: A True Story*, and she is featured in the collection *Owney, the Post Office Dog and Other Great Dog Stories*. Annie's gravesite is now a historical landmark. For several years there was an annual celebration, the "Annie Walk." Fifty years after her death, to note the beginning of the "Annie Walk," a small bronze statue of Annie's likeness was placed in front of the town library, with one paw held out as if saying, "Read, human! Read!"

Antis (Dog)

Called the "Dog Who Could Fly," Antis became famous during World War II for being the only mascot to fly in combat missions. Antis, a German shepherd puppy, was discovered in an old farmhouse by Czech gunner Robert Bozdech and French pilot Pierre Duval, where they sought shelter after being shot down by a German plane. They found the puppy emaciated and weak. Bozdech believed the little pup to be a fighter, and despite the risk, he tucked the little dog in his jacket, and the three of them made their escape. Later, the three were rescued by French soldiers. Duval had been injured in the crash and was taken to the hospital, but Bozdech and Antis flew back to the air base in Saint-Dizier, France, where the Czech airmen eagerly accepted Antis into their group. The men had originally named the dog Ant after the Russian ANT dive bombers but later renamed him Antis, so as

not to be confused with the English word "aunt." Once, Antis snuck aboard the plane and was discovered by Bozdech 20 minutes into the flight, and Bozdech shared his oxygen mask with the dog. Although it was against regulations for Antis to be aboard the plane, as Bozdech tried to make the case for Antis, his commanding officer interrupted, saying, "There's a very good English expression, 'What the eye doesn't see, the heart doesn't grieve over.' I believe it's more often used in connection to matters amorous, but it does just fine for last night's little escapade." On subsequent flights, Antis wore an oxygen mask made to fit him. On the ground, Antis also proved invaluable; he had the ability to detect enemy aircraft even before the planes could be seen or heard and also before they showed up on radar, saving many lives. He was also credited for saving five lives and a one-year-old child from the rubble of a building destroyed during a raid. In 1945, at the end of the war, Antis remained with Bozdech, who never had another dog after Antis. In 1949, Antis was awarded the Dickin Medal, the highest award given for animal bravery. Antis had been injured twice in action. In 1953, Antis died at the age of 13, and Bozdech died in 1980 at 67.

Apollo (Dog)

Apollo, a German shepherd in service with the New York Police Department beginning in 1994, was the first search and rescue dog on scene after the destruction of the World Trade Center in New York City on September 11, 2001—arriving within fifteen minutes. Around 350 search and rescue dogs that bravely risked their lives saving trapped people and later retrieving bodies were given recognition when Apollo was awarded the Dickin Medal on their behalf. Two guide dogs also received the Dickin Medal for rescuing their owners. Salty guided her owner, a Port Authority employee, down 70 flights of stairs in Tower 1. Roselle, trapped with her owner, a sales manager, led him down 78 flights of stairs after Tower 1 collapsed. Apollo died in 2006.

Aristides (Horse)

This small chestnut Thoroughbred unexpectedly won the first Kentucky Derby at Churchill Downs in Louisville on May 17, 1875. His owner, Hal Price McGrath, had expected his other horse, Chesapeake, to win. Aristides was entered to wear out the opposition with his early speed. However, Aristides took the lead and held it for the entire race. In 1875, the race was a mile and a half and remained at that distance until 1896, when it was changed to its present mile and a quarter. With odds of 29 to 10, Aristides won the race in two minutes, 37 seconds. Hal Price McGrath, along with trainer Ansel Williamson and jockey Oliver Lewis, won the smallest-ever Derby prize of $2,850. Aristides was foaled in 1872 and died on June 21, 1893.

Asta *see* Skippy

Athena (Owl)

In 1850, the British nurse Florence Nightingale visited Greece, and on June 5, while walking along the walls of the Acropolis, she noticed a group of children tormenting something fluffy, which turned out to be an owlet that had fallen from its nest. She chased the children away, rescuing the little owl. She named it Athena, after the Greek goddess of wisdom and war. After she cared for this nine-inch, brown-and-white speckled owl, they became inseparable friends. Florence Nightingale would carry Athena in her pocket and pull her out to distract her patients with the trained owl's bowing and curtseying. After five years, war broke out in Crimea, and Florence's nursing skills were needed in the field. During the time that she would be away, Athena was to live in her attic. Athena died in the attic, having been too domesticated to hunt for food herself, and from her extreme loneliness because family members neglected to check on her. Heartbroken, Florence had Athena preserved by a taxidermist. After Florence's death, Athena went on display at Claydon House in Buckinghamshire, the home of Florence's wealthy sister, Parthenope. Later, Athena ended up at an elderly care charity, Age Care, which also owned the Nightingales' childhood summer home, "Lea Hurst." When Athena and other artifacts went up for sale in the UK, a community fundraiser made enough money to purchase the owl, making it a permanent display at the Florence Nightingale Museum in London.

B2 (Tiger)

B2 was born in 1997 and until late 2011 he was the star attraction of India's Bandhavgarh National Park, where he was the dominant male. A glimpse of this endearing cat was a special gift. He would often play with his cubs in full view of cameras. Even local villagers were won over by this tiger, despite his habit of occasionally taking their livestock. He died in November 2011 from wounds inflicted upon him in a territorial dispute with a younger and stronger male tiger.

Babou (Ocelot)

Babou, a medium-sized American wildcat, was a furry companion to the Spanish surrealist artist Salvador Dalí during the 1960s. Dalí had his pet in tow wherever he went, including on the ocean liner SS *France* and also to an art gallery in Paris—infuriating the owner, who claimed that the animal was making a "nuisance" on some priceless engravings.

Dalí's response was "A nuisance of Dalí's can only increase their value." Not surprisingly, the owner raised the price of these works of art by 50 percent. Dalí also brought Babou with him to fancy restaurants. A funny anecdote describes an outraged woman who yelled at Dalí for bringing a wild animal to the restaurant. Dalí calmly explained that the ocelot was a house cat that he had painted to look "op art." It seems that Dalí had a fondness for bizarre pets. He had a pet bat as a child, and he was also photographed walking around Paris in 1969 with an anteater on a leash.

Baldy (Horse)

Baldy, a horse belonging to Civil War General George Meade, was very loyal and extremely strong both physically and mentally. Baldy had barely recovered from a severe wound he had sustained in the first battle of Bull Run when Meade purchased him. Meade and Baldy endured seven days of fighting in battles around Richmond in 1862. In the battle of Antietam, Baldy sustained a deep wound in his neck and was left for dead on the field. However, he was found alive and recovered so steadfastly that he went back into battle. He went on to participate in the battles of Chancellorsville, Fredericksburg, and Gettysburg. Near the end of fighting at Gettysburg, once again the horse sustained another bullet, which almost ended his life. By the time he was fully recovered, the war had ended. Baldy outlived Meade and followed the funeral procession to the grave, carrying the general's saddle with his boots reversed in the stirrups.

Baldy with his owner General George Meade in 1863 (Library of Congress).

Balto, a Siberian Husky, with his owner Gunter Kaasen, ca. 1925 (National Institutes of Health).

Balto (Dog)

Balto, a Siberian husky, was one of the most famous dogs in the world in the 1920s. Balto was among 150 sled dogs and mushers to trek 674 miles across Alaska in five and a half days to deliver a diphtheria antitoxin to Nome to treat a major epidemic among the Inuit. The teams faced hazardous conditions such as blizzards, whiteouts and extreme temperatures. Balto emerged a hero, leading the final sled dog team into Nome after saving his team in the Topko River. He was able to tirelessly stay on the trail even in the dark. Upon his arrival in Nome in the early morning of February 2, 1925, Balto's owner, Gunter Kaasen, said three words: "Damn fine dog." There is a statue erected in Balto's honor in New York City's Central Park. The annual Iditarod sled dog race started in 1973 to honor the serum run to Nome. Balto lived out his life at the Cleveland Metroparks Zoo.

Bamboo Harvester (Horse)

Bamboo Harvester was an American

Bamboo Harvester (aka Mr. Ed), the famous talking horse on the 1960s television series of the same name.

Saddlebred/part-Arabian horse foaled in 1949. He was famous for playing the hilarious talking horse Mister Ed and starring along with Alan Young on the television series of the same name, which ran from 1961 to 1966. Bamboo Harvester was trained by Les Hilton, who made the horse's lips move by pulling on a nylon thread placed in its mouth. According to Hilton and Alan Young, Bamboo Harvester made the connection that he needed to move his lips when Young quit talking. Cowboy star Allan "Rocky" Lane was the uncredited voice for Mister Ed. Bamboo Harvester was euthanized in 1970, at age 20–21, after suffering from kidney disease and arthritis.

Bamse (Dog)

Bamse, whose name is Norwegian for "cuddly bear," was a St. Bernard serving at sea for the Royal Norwegian Navy during World War II. His life at sea began on a Norwegian whale catcher, the *Thorodd*, under Captain Erling Hafto. At the beginning of the war, the *Thorodd* was converted to a coastal patrol vessel, and Bamse became one of the crew on February 9, 1940. On April 9, 1940, the Nazis invaded Norway, and the ship was used to transport POWs. During Bamse's service there arose many delightful, light stories as well as many stories of heroic acts. When fights would break out at the pub, he would subdue one of the soldiers by placing his large paws on his shoulders. Another account tells how Bamse rode a local bus to a nearby pub, unaccompanied, to retrieve and lead drunken sailors back to the ship. More importantly, Bamse performed many life-saving heroics, one of which included saving a lieutenant from being attacked by an enemy with a knife by pushing the attacker overboard. He also saved another sailor who had fallen into the cold water by leaping in and keeping the man afloat as they waited for help to arrive. After the war, Bamse became the mascot for the Free Norwegian Forces. Bamse died on July 22, 1944, from heart failure. He was buried with full military honors, and his coffin was draped with the Royal Norwegian flag. Eight hundred mourners—children, shopkeepers, factory workers, housewives, local dignitaries and the crews of six Norwegian ships—gathered for his funeral in Montrose. Bamse was buried in the sand dunes on the banks of the South Esk River. The Royal Norwegian Navy holds a commemorative ceremony every 10 years. Bamse also received many honors posthumously.

Bamse's story is documented by the Montrose Heritage Trust, a group operating a website devoted entirely to Bamse. The trust's website contains a series on Bamse's adventures, with testimonials from residents who remember him dearly.

Bandoola (Elephant)

Since ancient times, elephants have helped armies. Elephants have played an important role during both peace and war because of their size and strength, and also their intelligence. During World War II, Bandoola helped Lieutenant Colonel Jim Williams (nicknamed "Elephant Bill"), who was in charge of a British Army elephant company 700 strong in the jungles of Burma. Elephants were used for many tasks, such as pulling up trees for wood needed in England, moving logs to build bridges, launching ships, and carrying people and supplies across mountains, rivers and roads. In 1944, Bandoola led Williams' last 47 elephants to safety when the enemy was coming to take the elephants.

Bankes' Horse *see* Marocco

Barry der Menschenretter (Dog)

Barry der Menschenretter (the latter two parts meaning "people rescuer" in German) came from a breed that predated the modern Saint Bernard. The monks at the Saint Bernard Pass between Switzerland and Italy bred Saint Bernard dogs as far back as 1695. Barry lived at the Saint Bernard Hospice, a monastery in the Swiss Alps. Being a mountain rescue dog, he is credited with saving more than forty people over 12 years in the 1800s. Barry was capable of finding buried travelers, digging through the snow and lying on top of the injured to provide warmth until help arrived. Barry's most famous rescue was of a child who became trapped on an ice shelf. Barry reached and revived the boy, and when rescuers could not reach them Barry had the child climb onto his back so he could bring him to safety. Barry died in 1814 in Bern, Switzerland, where his body is preserved and exhibited in the Natural History Museum. There is also a monument honoring him in Paris. The hospice continues to honor Barry by always having one dog in residence named Barry.

Barry der Menchenretter, a mountain rescue dog in the 1800s.

Basket (Dog)

Basket, a white Standard Poodle, was bought by Gertrude Stein in 1928 at a dog show in Paris. He was named Basket by her partner, Alice B. Toklas, who felt he was so fashionable that he should carry a basket of flowers in his mouth. Ever since reading Henry James' *The Princess Casamassima*, Alice had always wanted a white poodle. Stein insisted that Basket be bathed in sulfur water each day and afterwards chased around outside to dry off. Basket died in 1937 and was followed by the second Basket, another white poodle. Pablo Picasso had urged Stein to get an Afghan hound.

Beachcomber (Pigeon)

Beachcomber was a Canadian Army carrier pigeon during World War II. These pigeons served a very important purpose so that soldiers, sailors, and pilots could communicate their progress, request supplies or call for help. Messages were written on small pieces of paper and placed in a small container attached to one of the pigeon's legs. Often two pigeons would be sent with the same message in case one didn't make it. When they landed, a bell or buzzer would go off, alerting the soldiers that a message had arrived. On March 6, 1944, Beachcomber was the only Canadian war pigeon to be awarded the Dickin Medal—for bringing the first news of the landing at the German-occupied port of Dieppe on the northern coast of France. The Dieppe Raid took place on August 19, 1942.

Beautiful Jim Key (Horse)

During the late 19th and early 20th centuries, a Quarter Horse named Beautiful

Basket pictured with his owner Gertrude Stein, June 13, 1934 (Henry W. and Alberta A. Berg Collection of English and American Literature, The New York Public Library).

Jim Key and his owner/trainer, Dr. William Key—a former slave, self-trained veterinarian and patent medicine salesman—became famous. Key emphasized that he used only patience and kindness in teaching Beautiful Jim. The horse became a celebrity, performing from 1897 to 1906, by reading and writing, making change with money, doing simple arithmetic, telling time, spelling, sorting mail and reciting Bible passages. Beautiful Jim Key performed at large venues from Atlantic City to Chicago, before an estimated 10 million Americans. He toured the United States in a special railroad car. Beautiful Jim Key's popularity carried the message of the importance of preventing cruelty to animals. More than two million children in America signed the Jim Key Band of Mercy, pledging, "I promise always to be kind to animals." The shows brought African Americans and whites together during a time when the two rarely interacted. Beautiful Jim had a "bodyguard" by the name of Monk, a stray dog. Beautiful Jim passed away in 1912. At an exposition in Tennessee, President William McKinley saw a performance and declared, "This is the most astonishing and entertaining exhibition I have ever witnessed."

Beautiful Joe (Dog)

Beautiful Joe's story brought awareness of animal cruelty to the world. His life was filled with abuse, beginning when he was just a puppy until he was rescued by the Moore family. His story, *Beautiful Joe*, was written by Canadian author Margaret Marshall Saunders after learning about his life from her sister-in-law, Louise Moore (named Laura Morris in the book). In order to make the greatest impact on readers, the story was written from the dog's

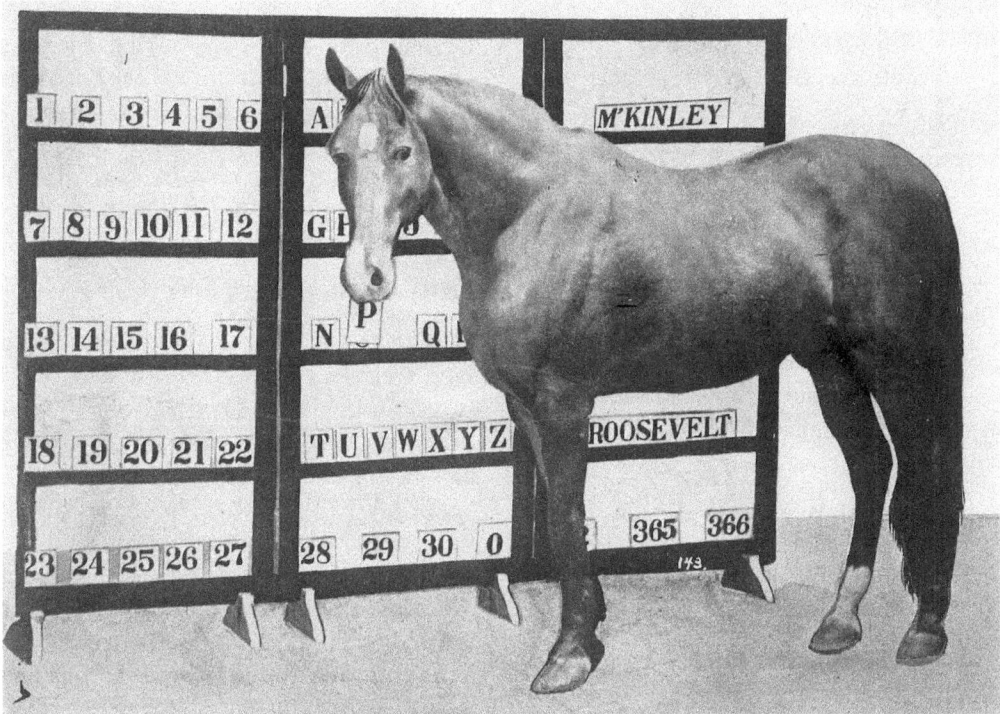

Beautiful Jim Key at the 1904 World's Fair (Missouri History Museum).

point of view. Born in Meaford, Ontario, Canada, Joe belonged to a man named Jenkins (referred to as the "bad man" in the book), who was unbelievably cruel to his animals by neglecting, starving, and beating them. Joe tells about the day when, at eight weeks old, he saw his brothers and sisters put to death, one after another, in front of their mother, who died "of a broken heart" shortly afterward at only four years old. Joe says of that day, "My mother ran up and down the stable screaming with pain." For some reason, Joe was spared. There came a day when Joe could take no more—when Jenkins kicked him and Joe bit the man. Unfortunately, the man was furious and cut off Joe's ears and tail with a hatchet. As Joe describes in the book, "The pain all through my body was dreadful. My head seemed to be on fire, and there were sharp, darting pains up and down my backbone." Through the nurturing of the Moore family, he lived the rest of his life surrounded with love among friends.

The story was published in 1893 and by the late 1930s had sold over seven million copies. (The book can be found online at: digital.library.upenn.edu/women/saunders/joe/joe.html.) To honor Beautiful Joe's life and the achievements of Margaret Marshall Saunders, the Beautiful Joe Heritage Society was formed in 1994. This famous dog was laid to rest in Beautiful Joe Park in Meaford.

Beauty (Dog)

Beauty, a wire-haired terrier born on January 4, 1939, was a search and rescue dog during World War II. She was owned by Bill Barnet, superintendent for the People's Dispensary for Sick Animals (PDSA) in the UK. One day in 1940 while out on a rescue mission with her owner, she began digging in the rubble of a bombed building. Within minutes a cat emerged; in all, Beauty rescued no less than 63 animals during her war service. Beauty died on October 17, 1950, and was buried in the PDSA Ilford Animal Cemetery.

Beauty's service awards include:

PDSA Pioneer Medal

A silver medal inscribed, "For Services Rendered" and awarded by the deputy mayor of Hendon, a London suburb

Freedom of Holland Park

The Dickin Medal on January 12, 1945, inscribed, "For being the pioneer dog in locating buried air-raid victims while serving with a PDSA Rescue Squad"

Her medals were on display in the Animals' War Exhibition at the Imperial War Museum North, which ran from May 26, 2007, to January 6, 2008.

Beauty (Dog)

When magicians Lafayette and Harry Houdini were performing at the Grand Opera House on Cherry Street in Nashville, Houdini gave his friend Lafayette a terrier dog that would become his best friend, traveling companion and also a member of his show. Lafayette gave Beauty her own bedroom and bathroom in his house, served her five-course meals and had her wear a diamond-studded collar. On the front door of his home, he hung a plaque reading, "The more I see of man, the more I love my dog." Lafayette was the highest-paid entertainer in Vaudeville due to a fabulous show. Some of his famous acts included "The Lion's Bride" in which supposedly a woman was placed in a cage with a ferocious lion, and when the lion prepared to pounce, it was revealed to be Lafayette in a lion's costume. In one routine with Beauty, Lafayette dressed in an artist's smock and beret and began to paint until he ripped the paper from the frame and Beauty burst out. Tragedy struck for Lafayette when on May 5, 1911, Beauty passed away. Beauty was buried in the Piershill

Cemetery in Edinburgh, Scotland, where Lafayette too would be buried not long after. On May 9, 1911, a fire broke out at the Empire Theatre in Edinburgh. Lafayette perished with ten others. On May 14, 1911, Lafayette's body was cremated and the urn was buried between Beauty's paws.

Belka and Strelka (Dogs)

Belka and Strelka (their names mean "squirrel" and "arrow" in Russian) were two female mixed Samoyed husky dogs that were the first animals to survive orbital flight. The dogs were launched on *Sputnik* V by the Soviet Union on August 19, 1960, and accompanied by a rabbit, some rats, many mice, and an assortment of insects and plants. In twenty-five hours they completed just over seventeen orbits and returned safely. Later both dogs had normal litters, and one of Strelka's puppies, Pushinka (name means "fluffy" in Russian), was given to Caroline Kennedy as a gift when Prime Minister Nikita Khrushchev visited the United States in 1961.

Benji *see* Higgins

Betsy (Dog)

Betsy is a black and white long-haired border collie born in 2002 who lived up to the belief that border collies are considered the most intelligent dogs in the world. Betsy is cited as being one of the smartest animals in history. At 10 weeks old she could sit on command and also retrieve objects by name. It has been said that Betsy's "understanding of human forms of communication is something new that has evolved. It is something that has developed in the dogs because of their long association with humans. This is not trivial, but it means that evolution can invent similar forms of advanced intelligence more than once. It is not something reserved only for primates or mammals" (Listverse website). Betsy learned in a similar way to toddlers. One of her feats included the ability to decipher a similarity between a two-dimensional photograph and its object; without training she could find an object after being shown an image. She could also identify 15 people by their name. Scientific researchers have recognized some links between human and dog diseases. In 2009, they identified a similarity between Lou Gehrig's disease and a genetic mutation for degenerative myelopathy in dogs. Also in 2011, the same gene mutation found in Tibetan terrier dogs was related to a neurological disorder related to Parkinson's disease. It is hoped that perhaps possible treatments for cancers in humans can be found by studying genetic causes of cancers in dogs.

Bill the Bastard (Horse)

Bill the Bastard, one of 130,000 Australian horses that served in World War I, became a Great War legend. In a time when military horses were treated very brutally, his rider, Michael Shanahan, was never brutal with him and gave Bill the chance to become a war hero. Shanahan was given a Distinguished Service Order and Bill went down in history for his incredible stamina when he carried four Tasmanian troopers on his back for six hours. Even after retirement, Bill would carry machine guns or lead the line. Bill's story has a happy ending; he was left with villagers on Turkey's Gallipoli Peninsula, the resting place of Australia's fallen war heroes.

Billy (Pygmy Hippopotamus)

Billy (AKA William Johnson Hippopotamus) was captured in Liberia in 1927 by Firestone Tire & Rubber Company founder Harvey Samuel Firestone, who in turn gave Billy to U.S. president Calvin Coolidge. The president had a reputation

Billy was a gift to President Calvin Coolidge from Harvey Samuel Firestone.

for his collection of animals, many of which (including Billy) he donated to the Smithsonian's National Zoo in Washington, D.C. Since pygmy hippos did not exist in the United States at the time, Billy is considered the common ancestor to most of the pygmy hippos in American zoos. Billy was a very popular and valuable animal at the zoo, being only the eighth pygmy hippopotamus brought to the United States. *The New York Times* wrote that Billy was "as frisky as a dog. Even the antics of the monkeys go unobserved when the keeper opens the tiny hippo's cage and cuts up with him." On September 4, 1929, a female pygmy named Hannah was acquired by the zoo to be Billy's mate. Between 1931 and 1954 Billy and Hannah had 15 calves, with seven living at least one year. In 1940 the zoo acquired a second mate for Billy named Matilda, and between 1943 and 1956 Matilda gave birth to eight calves, six being reared. The zoo began naming all of Billy's calves Gumdrop followed by Roman numerals. Billy died on October 11, 1955, and Hannah died on March 6, 1958.

Bing (Dog)

Bing (AKA Brian) the paradog, a two-year-old Alsatian-collie, was one of the first dogs to join his fellow British paratroopers to parachute behind enemy lines in World War II. Lance Corporal Ken Bailey was assigned to run the War Dog Training School in Hertfordshire. As the 13th (Lancashire) Parachute Battalion prepared for D-Day, they launched an experimental program to enlist dogs. These "paradogs" (short for "parachuting dogs") were trained to locate mines, keep watch and warn of the presence of the enemy. In 1941, the British War Office made an appeal to dog owners to loan their dogs to the war effort. Due to rationing, Bing's owner, Betty Fetch, gave Brian (his civilian name) to the cause. Bailey also trained

two other Alsatians, Monty and Ranee, to be a part of Britain's paradogs program. Over two months the dogs were trained to get used to loud noises, sit for hours on transport aircraft with the propellers whirring, recognize the smells of explosives and gunpowder, and become familiar with battle scenarios—what action to take if their handler was captured, how to track down the enemy, and how to behave during firefights. The paradogs were taught parachuting maneuvers. The training jumps seemed to go well, with the dogs allowing themselves to be thrown out of the planes or willingly leaping out on their own. Bing and his handler first saw action in a Recce Platoon with the 13th Parachute Battalion, part of the 6th Airborne Division. Just before D-Day, June 6, 1944, three planes, each holding 20 men and 1 dog, took off at 11:30 p.m. on June 5 and arrived at Normandy at 1:10 a.m. Bing's jump did not go smoothly. His parachute got caught on some tree branches, and he waited for two hours before being found. Bing and the other dogs proved to be very good at locating mines and booby-traps and sniffing out the enemy, saving many lives. Monty was severely wounded and Ranee was separated from her battalion and was never seen again. On March 24, 1945, along with his new handler, Corporal "Jack" Walton, Bing was dropped over the Rhine while taking part in Operation Varsity and Operation Plunder, and advanced into Germany. Surviving the war, Bing was awarded the Dickin Medal, the UK's highest honor for animals, and he was returned to the Fetch family. When Bing died in 1955 he was buried in a cemetery of honor for animals. A life-size statue of Bing wearing his parachute is located at the Museum of the Parachute Regiment and Airborne Forces in Duxford; next to his medal are the words "For Gallantry" and "We Also Serve."

Bingo (Dog)

This black and white dog and Sailor Jack have appeared on Cracker Jack packages since 1919. The model for Jack was Robert Rueckheim, whose grandfather founded a popcorn business in 1873. Bingo was Robert's dog.

Binti Jua (Gorilla)

Binti Jua (name means "daughter of sunshine" in Swahili) was a female western lowland gorilla born on March 17, 1988, who lived in the Brookfield Zoo in Illinois. Binti came to the attention of international media when on August 16, 1996, a three-year-old boy fell from a wall 24 feet into the gorilla enclosure. Binti cradled the unconscious boy and laid him by an access door, where he was rescued by zoo staff. After four days in the hospital with a broken hand and a gash on his face, he made a full recovery. In the aftermath of the incident, experts debated if Binti's actions were from training at the zoo or animal altruism. However, primatologist Frans de Waal believes Binti's actions were a result of empathy. There are other examples of animals demonstrating altruism. Jambo, a male gorilla in the Jersey Zoo, saved a five-year-old child who had fallen into his enclosure. Also, chimps have been observed comforting each other after trauma.

Black Diamond (Bison)

Black Diamond (AKA Toby), a North American bison, was born in 1893 and given to the Central Park Menagerie (now the Central Park Zoo) by Barnum and Bailey. Another version is that he was a circus animal retired to the Central Park Zoo from the Barnum and Bailey Circus. Legend has it that in 1911 American sculptor James Earle Fraser used Black Diamond as the model for the U.S. buffalo nickel coin introduced in 1913. In his book

Black Diamond may have been the model for the U.S. buffalo nickel introduced in 1913.

Renaissance of American Coinage 1909–1915, Roger W. Burdette states that there is some uncertainty that Black Diamond was the model because Fraser also made sketches at the Bronx Zoo, and there are no records of Black Diamond living at the Bronx Zoo in that era. Black Diamond may also have been the model for the face of the $10 U.S. Banknote, Series 1901. On June 28, 1915, at age 22 Black Diamond was put up for auction, but there were no bids. Game and poultry dealer A. Silz, Inc., purchased Black Diamond for $300 and he was slaughtered on November 17. "Black Diamond Steaks" sold for $2 per pound.

Black Diamond (Elephant)

Black Diamond, an enormous Indian elephant weighing 18,000 pounds, was owned by the Al G. Barnes Circus. He was prone to having rages, so when he was displayed in public, he was kept chained to a couple of female elephants to keep him calm. On October 12, 1929, in Corsicana, Texas, in a rage he injured a former trainer and killed a woman; it was at this point the circus decided he was too dangerous to keep in the show. They initially tried to poison him and when that failed, Hans Nagel, the keeper at the Houston Zoo, shot Black Diamond to death with at least 60 bullets.

Black Jack (Horse)

This coal-black Morgan–American Quarter Horse crossbreed served in the Caisson Platoon of the 3rd U.S. Infantry Regiment. He was named in honor of General John J. ("Black Jack") Pershing. Black Jack was the riderless horse in more than 1,000 U.S. Armed Forces Full Honors Funerals, the majority held in Arlington National Cemetery. Black Jack was foaled on January 19, 1947, and entered the military in 1953. The empty-saddled (riderless) horse rode with a deceased rider's boots reversed in the stirrups to symbolize a fallen hero. Black Jack had a long military career, participating in four state funerals: presidents John F. Kennedy (1963), Herbert Hoover (1964), Lyndon B. Johnson (1973), and General Douglas MacArthur (1964). Black Jack retired in 1973 and died in 1976 after a 29-year military career. He was buried with full military honors after being cremated. The tradition of the riderless horse traces back to the ancient Mongols. When a warrior was killed, his caparisoned horse would be led to the burial and then sacrificed.

Blackie (Dog)

Blackie (Brand Number H24) and his handler, Corporal Technician Kido, went on patrol on April 12–13, 1945, with Company F, 123rd Infantry. The mission was successful in locating a temporary encampment of 500 Japanese without being

Black Jack served in the Caisson Platoon of the 3rd U.S. Infantry Regiment as a riderless horse in the funeral for President John F. Kennedy, November 25, 1963 (David S. Schwartz, U.S. Army, John F. Kennedy Library and Museum).

detected. It was a reconnaissance mission so there was no contact with the enemy.

Blackie Halligan (Pigeon)

Blackie was hatched at Fort Monmouth, New Jersey, and later shipped to Guadalcanal to serve with the 132nd Infantry Regiment, 33rd Infantry Division. Blackie carried many different messages on short missions. However, one day he was carrying a message detailing the location of 300 Japanese troops when he was shot down; amazingly, showing his dedication, he successfully delivered the message five hours later, on a trip that should have taken only twenty minutes. He was decorated for bravery under fire by General Alexander Patch and returned to the United States after the war.

Blackwall and Poplar (Cats)

In 1901, Robert Falcon Scott set sail for the Antarctic in the brand-new ship *Discovery*; he carried a crew of 48 men and two cats, Blackwall and Poplar. Poplar, a tabby and white tomcat, quickly befriended Scott and spent most nights on his bunk along with Scamp, the captain's black Aberdeen terrier. Blackwall was a black female cat that became American stoker Arthur Quartley's pet. In early December 1902, Blackwall gave birth to one small kitten, and in mid–February 1903 she gave birth to six kittens. The *Discovery* spent the winters of 1902 and 1903 stuck in the Antarctic ice until being rescued in early 1904 by two ships, the *Terra Nova* and the *Morning*, using dynamite to break up the ice. The diary entries of Lieutenant Charles Royds shed some light on Poplar's fate. On March 22, 1903, Poplar was put to death by being dropped down a seal hole by unanimous agreement. Apparently, the crew felt that Poplar was a cat with dirty habits. On February 2, 1904, having some spite against Arthur

Quartley, seaman Whitfield set two dogs loose on Blackwall and she was killed. The *Discovery* is berthed at Dundee in Scotland and is open to the public.

Blair (Dog)

Blair, the world's first canine movie star, made his debut in 1905 in *Rescued by Rover*, a six-and-a-half-minute film of a collie rescuing his master's baby from kidnappers. British film pioneer Cecil Hepworth produced the movie, which was directed by Lewin Fitzhamon and featured Hepworth's wife, baby, and dog named Blair. The demand for copies of the film was so great that Hepworth had to reshoot it twice because the negatives had fallen apart. More successful sequels were made before Blair died. Nearly all of Blair's films have been lost. In the few films that still exist, he can be identified by a diamond mark on his forehead and a distinct white ruff on his left side and sable on his right side. Even before the Rover films, Blair had been Hepworth's faithful and constant companion and the general pet of the studio. It was after these films that the uncommon name of Rover became one of the most recognized names for dogs in the English-speaking world.

Blind Tom (Horse)

Blind Tom, appropriately named for being blind, played an integral part in building the world's first transcontinental railroad in the 1860s, which in turn helped to build the West. Despite being blind, he was chosen to be the lead horse for the Union Pacific line, pulling the flatcars to the end of the trackline. He was known for his great strength—pulling 110,000 tons of rails in just a few years—and his tenacity in struggling through snow, mud and cold. During the building of the railroad he earned a certain celebrity status, being mentioned in several newspapers. In 1869, when the railroad was finished, he was the only horse mentioned. Twenty-seven months later, he was working in Omaha, Nebraska, and his blindness still did not interfere with his work. Blind Tom may or may not have existed. There is no primary historical source, but there is some mention of a Champion Tom pulling carts of iron.

Blue Boy (Dog)

Blue Boy's handler, Matheson, describes the amazing job the Bedlington terrier did delivering his message at Mount Kemmel under intense machine-gun fire during the Great War.

The Blue Hen (Pigeon)

The Blue Hen is not an official name for this pigeon. This is an account of a blue hen that traveled twenty-two miles in just twenty-two minutes, when most pigeons can only fly about forty-five miles per hour. This particular blue hen was released from a seaplane off the Scottish coast. After their plane went down twenty-two miles from base, the rescued men learned of this hen's incredible flight.

Boatswain (Dog)

Boatswain won the heart of Lord George Gordon Byron from the time this Newfoundland dog entered the Byron home in 1803. The poet and pup were never apart and always getting into trouble. Boatswain developed rabies after being bitten by an infected dog. Byron stayed with him until the dog's death on November 10, 1808. The poet honored Boatswain by erecting a large monument on his family estate. The tomb is engraved with a poem written by Byron, "Epitaph to a Dog."

"Epitaph to a Dog"
By George Gordon Byron,
1788–1824

Near this Spot
are deposited the Remains of one
who possessed Beauty without Vanity,
Strength without Insolence,
Courage without Ferosity,
and all the virtues of Man without his
Vices.

This praise, which would be unmeaning
Flattery
if inscribed over human Ashes,
is but a just tribute to the Memory of
Boatswain, a Dog
who was born in *Newfoundland* May
1803
and died at *Newstead* Nov. 18th, 1808.

When some proud Son of Man returns to
Earth,
Unknown to Glory but upheld by Birth,
The sculptor's art exhausts the pomp of woe,
And storied urns record who rests below:
When all is done, upon the Tomb is seen,
Not what he was, but what he should
have been.
But the poor Dog, in life the firmest friend,
The first to welcome, foremost to defend,
Whose honest heart is still his Master's
own,
Who labours, fights, lives, breathes for
him alone,
Unhonoured falls, unnoticed all his worth,
Denied in heaven the Soul he held on
earth-
While man, vain insect! hopes to be
forgiven,
And claims himself a sole exclusive
heaven.

Oh man! thou feeble tenant of an hour,
Debased by slavery, or corrupt by power-
Who knows thee well, must quit thee with
disgust,
Degraded mass of animated dust!
Thy love is lust, thy friendship all a cheat,
Thy tongue hypocrisy, thy heart deceit!
By nature vile, ennobled but by name,
Each kindred brute might bid thee blush
for shame.
Ye! who behold perchance this simple
urn,
Pass on—it honours none you wish to
mourn.
To mark a friend's remains these stones
arise;
I never knew but one—and here he lies.

As a pet owner, poet Lord Byron kept an animal menagerie throughout his life, which included: A half-tamed wolf called

Lyon, a fully-grown bear that he brought to Cambridge University, 10 horses, 8 dogs, 3 monkeys, 5 cats, 5 peacocks, an eagle, a falcon, and an Egyptian crane.

Bob (Cat)

James Bowen went from being a recovering heroin addict living on the streets and making his living as a street musician in 2007 to becoming a "celebrity" with the once stray cat, Bob, in just ten years. James met Bob one evening in March when he found an injured and hungry ginger tomcat in his stairwell. Given his circumstances James did not feel he could take the cat in, but when 36 hours later the cat was still there, he gave him some food. When no one in the neighborhood claimed the cat, James took him to the RSPCA's Harmsworth Animal Hospital to be treated for his injuries. He named the cat Bob after a character on *Twin Peaks*, and eventually the two began traveling the streets of London while James played his music. In 2008, James began selling the *Big Issue*, a weekly newspaper bought by the disadvantaged and sold at a profit. Literary agent Mary Pachnos spotted James and Bob's unique story, and in 2012, *A Street Cat Named Bob* was published by Hodder & Stoughton. A second book, *The World According to Bob*, was published in 2013. Book signings became very popular and the two also made appearances on national TV in the United Kingdom. In April 2013, Bob was awarded "The Tails of the Unexpected Honour," a special award given to exceptional animals. In 2013, James chose the Blue Cross as the charity for him and Bob to support. In 2013, James and Bob's time was devoted to book signings and charity work, raising $33,000 for the Blue Cross. There have also been four children's books, a picture book spin-off from the blog *Around the World in 80 Bobs*, and a short story, *A Gift from Bob*. Bowen's books have sold over seven million copies worldwide. In 2016 a film about James and Bob's lives opened to good reviews.

Bob (Dog)

Bob the Railway Dog (also known as "Terowie" Bob), a stray thought to be a cross between a German Koolie and a Smithfield, left a legacy as a big part of South Australian Railways folklore. Towards the end of the 19th century, Bob was very well-known to the railwaymen for his occupation of Railway Traveler for the South Australian Railways. Seth Ferry, a special guard at Petersburg (present day Peterborough), acquired Bob on September 24, 1884. Bob traveled thousands of miles to and from Petersburg. According to the *Petersburg Times*, "His favourite place [was] on a Yankee engine; the big whistle and belching smokestack seem[ed] to have an irresistible attraction for him … he lived on the fat of the land, and was not particular from whom he accepted his dinner." A commercial traveler took a liking to Bob and bought him a special collar.

Two brass plates were inscribed with: "Stop me not, but let me jog, for I am Bob, the driver's dog"

Presented by McLean Bros. & Rigg

Bob died on July 29, 1895, and was eulogized around the world. His collar, photographs and other artifacts are displayed at the National Railway Museum in Port Adelaide. In November 2009, a statue of Bob was unveiled on the eastern end of Main Street.

Bob (Dog)

Bob, a mixed breed, was the first dog to receive the Dickin Medal, in March 1944. He was traveling with his unit, the 6th Battalion Queen's Own Royal West Kent Regiment, in Green Hill, North Africa, when he suddenly froze and refused to move. Bob was trying to forewarn the unit that the enemy was close by. Because of Bob's actions, the soldiers in the unit remained unharmed.

Bob, railway traveler for the South Australian Railways, sits on top of a locomotive at Port Augusta Railway Yard, ca. 1887 (State Library of South Australia, Port Augusta Collection, B 6422).

Bobbie (Dog)

"Bobbie the Wonder Dog," a two-year-old Scotch collie–English shepherd mix, rose to fame after his six-month ordeal, which began in 1923 while on a family road trip from Silverton, Oregon, to Indiana. In Indiana, Bobbie got separated from his owners, Mr. and Mrs. Frank Brazier. After searching and searching they failed to find him and, broken-hearted, returned to Oregon thinking he was lost to them forever. Miraculously, six months later, in February 1924, Bobbie turned up at the Braziers' doorstep looking mangy and scrawny and with his feet worn to the bone; it appeared as if he had walked the entire way back home. During the six months, he had traveled 2,551 miles (approximately 14 miles per day) across plains, desert, and mountains in the winter. After the local newspaper, *The Silverton Appeal*, published Bobbie's story it quickly spread across the country, and he was dubbed "Bobbie the Wonder Dog."

Bobbie received hundreds of letters from around the world and was honored with medals, keys to cities, and a jewel-studded harness and collar. He was featured in *Ripley's Believe It or Not!* and played himself in a 1924 silent film, *The Call of the West*. Bobbie died in 1927 at the age of six and was buried at the Oregon Humane Society's pet cemetery in Portland. A week later, Rin Tin Tin laid a wreath at his grave. Silverton's annual children's pet parade was started several years after Bobbie's death to honor the pets in our lives.

Bobby (Dog)

Bobby, a sheepdog mix, was adopted by the Royal Sussex Regiment after being captured with German prisoners in 1915. Bobby was a messenger dog and disappeared during a heavy battle in the trenches. In October 1916, Bobby was found in northern France in a chateau that had been converted into a field hospital. Bobby served another purpose in April 1917. The hospital was bombed and the wounded soldiers were moved to the cellar of the chateau, which was infested with rats. Bobby caught and killed all of the rats, keeping the cellars rat-free for the duration that the soldiers were held there. Bobby had been brought to the field hospital by a nurse who had found him wandering as if he were lost. After a while he resumed his work as a messenger dog for his regiment, and after the war he was adopted and returned to England, where he lived out a happy life.

Bobo (Dog)

During World War II, Bobo and his handler Sergeant John Coleman led a patrol into German-held territory. With the mission accomplished, while they were heading back to their outpost, Bobo alerted the patrol of a German patrol beginning to surround the outpost. A scout was sent ahead to warn their fellow soldiers; the Germans were dispersed and the soldiers went on to headquarters.

Bonfire (Horse)

Bonfire was one of millions of horses carrying men into battle, delivering supplies, and pulling equipment through fields. The chestnut gelding was a gift from Dr. John Todd to his friend Dr. John McCrae, who was stationed at Camp Valcartier near Quebec City, where Canadian soldiers were being trained before being shipped overseas. On January 24, 1918, Dr. McCrae died of both pneumonia and meningitis. John Almond, director of chaplain services, wrote of McCrae's funeral, "I have never seen such a gathering of military nobilities before in France. It was a great honor and well-deserved tribute the honor and esteem in which Colonel McCrae was held." Bonfire followed the casket with his bridle laced with white ribbons and McCrae's boots placed in the stirrups. Bonfire's fate following McCrae's death is unknown. One theory is that he was sold and retired in northern France. Another theory is that he was given to Lieutenant Colonel Bartlett McLennan, 42nd Battalion, Canadian Black Watch. Unfortunately, McLennan died at the end of the war and if Bonfire was indeed with him, his fate remains unknown. McCrae was so smitten with Bonfire he wrote to his sister, "I wish you could meet [Bonfire], he is one of the dearest things in horses that one could find ... he puts up his lips to your face and gives a kind of foolish waffle of his lower lip that is quite comical." Susan Raby-Dunne's 2012 book, *Bonfire— The Chestnut Gentleman: The True Story of John McCrae's Journey through WWI, as Told by His War Horse, Bonfire*, is not only based on true events of the war, but also tells of the events leading to McCrae's famous poem "In Flanders Fields."

Boomer Jack (Dog)

Boomer Jack, a mixed-breed dog, appeared in 1910 using the Northwestern Pacific Railroad system as his home. He rode the rails between Trinidad and San Francisco Bay, and at one point rode cross-country to South Carolina and back. The men of the Northwestern Pacific fed and cared for Boomer Jack because of his sense of independence and freedom. He never overstayed his welcome, traveling only every couple of days. At one point, Boomer Jack vanished but somehow made his way to South Carolina. The railroad men received a telegram asking about a dog with an NWP badge on his collar. Boomer Jack had suffered a severe leg injury from falling off a train but had continued his journeys. A bank account was set up for him from donations to help pay his medical bills. His lame leg did slow him up, and in 1926 Boomer Jack was found dead and was buried in a small redwood coffin at the switchyard.

Bouee (Dog)

Bouee was a small yellow and black dog without a tail and known for his regularity and quickness. He was cited on three occasions; his third citation was for his service as a messenger dog. All telephone connections had been destroyed, and under heavy fire he was able to dodge the shell explosions and continue to his destination.

Boxer (Dog)

Boxer was an Airedale assigned to carry messages along with Flash the Lurcher for the 34th Division. Both dogs were very fast and could deliver a message through very deep mud for four miles in as little as twenty minutes, whereas a man would take two hours. Dixon, Boxer's handler, reported that one day Boxer's run was slower than usual because he had stopped to eat a carcass in the field, and when Boxer returned, he slipped into his bed knowing that he had done wrong.

Brian *see* Bing

Brighty (Burro)

Brighty lived from about 1882 to 1922 and was first seen in the Grand Canyon near an abandoned miner's tent, sitting vigil. Brighty led a hybrid existence—not exactly wild, but not domesticated enough to be useful. Brighty was an independent little burro known as the "hermit of Bright Angel Creek." If caught, he would struggle free, but he was friendly and willingly carried supplies for some engineers or hauled water for the National Park Service. Marguerite Henry immortalized the burro in the children's novel *Brighty of the Grand Canyon* (1953). It was along the North Rim that early canyon tourists first met Brighty, probably between 1917 and 1922. Brighty was eaten in the winter of 1922 by two fugitives stranded at the Grand Canyon during a winter storm. The burro was not native to Arizona, having arrived with the Spaniards. It was decided to restore the canyon to a pre–Columbian state. Rather than relocating the burros, in 1924 rangers began hunting them. The hunts went on with little public knowledge, until the 1953 book and a 1967 Disney movie of the same title helped launch a campaign to save the remaining canyon burros. Rangers were still hunting burros even in the 1960s using helicopters. Money was raised in the 1980s to airlift rescue some of the remaining burros. There is a life-size bronze statue in the lobby of the Grand Canyon Lodge on the North Rim.

Marguerite Henry's comment on Brighty's statue:

> "The artist
> captured the soul
> of Brighty,
> forever wild,
> forever free."

Brownie (Dog)

Brownie was famous for being the Town Dog of Daytona Beach on Florida's Atlantic Coast. In 1939, a tan puppy was born and had no home until he met Ed Budgen, owner of the Daytona Cab Company, who adopted him. Budgen and his drivers named him Brownie and built him a dog house outside of their business. Soon the townspeople became very fond of Brownie for all that he did for the town. During the day he played with the children; often at night he went on patrol with the police officers; he had his picture taken by tourists; he kept people company while they were waiting for a cab or bus; and he would lift the spirits of people feeling sad. In a 2016 interview, Fayn LeVeille who played with Brownie as a child, was quoted in *The Daytona Beach News-Journal*, describing Brownie as "a very special animal, one that you might only meet once in a lifetime and that when he looked into your eyes it was as if he could see right through to your soul." It seems Brownie was so well-loved that locals and tourists would donate money in a coin jar by his doghouse to help pay for food, a dog license, and vet bills for incidents such as when he was hit by a car and got a broken hip. Brownie's story reached many around the country as it appeared in national newspapers and magazines; every year his fans sent him Christmas cards and gifts. Brownie died of old age on October 31, 1954, and at his funeral Mayor Jack Tamm said, "Brownie was indeed a good dog." He was buried in Riverfront Park under a slab of granite with an inscription reading, "The Town Dog, A Good Dog."

Bruce (Dog)

On February 17, 1945, Bruce bravely attacked three Japanese soldiers advancing with bayonets to a foxhole with two wounded American soldiers, saving their lives.

Bruin (Bear)

Soldiers in Company E of the 12th Wisconsin Infantry Regiment had a young bear as a mascot. Even though it was for a short time, he made such an impact that a soldier of the regiment wrote an entire book chapter on this uncommon companion. Bruin's owner, 16-year-old Harlan Squires, had enlisted with his pet in 1861. The company was pleased with the two new recruits and made modifications to house the bear. People came to see the bear. When the regiment left Wisconsin via train, Bruin traveled with the soldiers to Chicago, where he participated in a parade throughout the city. The train continued on to Missouri and eventually Leavenworth. Wherever Bruin went, he continued to amuse people. When Squires

learned his regiment was traveling to Texas or New Mexico, he sold Bruin to a citizen of Leavenworth for seventeen dollars.

Brumas (Polar Bear)

Brumas, born on November 27, 1949, and reared from mother Ivy and father Mischa, was the first baby polar bear born in Britain. She is named after her keepers at the London Zoo, Bruce and Sam (whose spelling is reversed in Brumas' name: M-A-S). Because Brumas was such an immediate attraction, the zoo's attendance broke the three million mark in 1950 and has not been exceeded since. Brumas was a female; however, the press reported that the bear was a "he," which was not corrected, so the public thought the bear was a male. Brumas died on May 17, 1958.

Bubbles (Chimpanzee)

Bubbles was born in early 1983 in an Austin, Texas, research facility that bred primates for animal testing. There are conflicting reports as to how this eight-month-old chimp became part of Michael Jackson's life. Bubbles was kept at the Jackson family's Encino home until 1988, when he was moved to Jackson's Neverland Ranch. Bubbles slept in a crib in Jackson's bedroom, ate candy in the Neverland movie theater, was fed at the dining table, wore a diaper, and used Jackson's personal toilet. The relationship grew as Bubbles began accompanying Michael Jackson on many outings and became his best friend. Michael Jackson had asked doctors if it was possible to give him an operation to alter his vocal cords. By 2003, Bubbles was becoming difficult to handle and was moved to an animal sanctuary. Since that facility closed in 2004, Bubbles has been kept at the Center for Great Apes in Florida. He loves to paint and some of his work has sold. He also enjoys playing with water and listening to the flute.

Bubi (Eagle-Owl)

On June 6, 2007, when the Finland national football team was playing against Belgium, the game was stalled for several minutes when an eagle-owl flew down, perching on the crossbar of the goalpost. Finland won the game; Bubi became their mascot and the team was named "Huuh-kajat" (the Finnish word for "eagle-owls"). Bubi was given the prestigious "Helsinki Resident of the Year" prize by an association of journalists. Pasi Tuohimaa, the head of the association, said, "Bubi promoted Finnish football much more efficiently than our team's scoreless draws here and there."

Bucephalus (Horse)

Prince Alexander the Great, son of King Philip II, was twelve years old when he met Bucephalus (the name means "oxhead"), the unruly and headstrong black horse with a white star on his forehead. To everyone's amazement, Prince Alexander was able to tame the horse by discovering that the animal was afraid of his own shadow because he thought he was being challenged by a huge black horse. By turning the horse away from the sun so that he could not see his shadow, Prince Alexander was able to soothe him enough so that he was tamed. Being inseparable, Prince Alexander and Bucephalus fought in many battles and won victories from Macedonia to India. There is some dispute as to the cause of the horse's death. One theory is that he died of fatal injuries in the Battle of the Hydaspes in 326 BCE. The other theory is that he simply died of old age. Alexander founded a city, Bucephalus, in honor of his horse. It is believed that the horse was buried in the town of Jalalpur Sharif.

Bud Nelson (Dog)

Bud's story actually begins on May 23, 1903, when 31-year-old Horatio Nelson Jackson, automobile enthusiast and former physician, and 22-year-old Sewall Crocker, a gasoline engine mechanic and former bicycle racer, left San Francisco to make the first cross-country drive to New York City on a $50 wager. Roads in 1903 were very unkempt, with no road signs, numbers or gas stations. Most attempted cross-country trips failed. After 19 days (which today would take about 13 hours), they rolled into Idaho in their cherry red Winton touring car. In Idaho Jackson got an offer to buy a young, light-colored bull terrier for $15 (today it would cost $400); he took the offer and named his new traveling companion Bud. To protect his eyes from dust, Bud was fitted with goggles and became a big attraction everywhere they went. Despite many breakdowns, waiting for car parts, getting lost and having an accident, they drove into New York City on July 26. The 4,500-mile trip had taken 63 days, 12 hours and 30 minutes, and 800

Bud, Horatio Nelson Jackson's travelling companion on the first cross-country drive from San Francisco to New York City.

gallons of gasoline. Bud lived out a happy life with Jackson in Vermont. The car and Bud's goggles were donated to the National Museum of American History. Jackson was quoted saying that Bud was the only one who "used no profanity for the entire trip."

Buddy (Dog)

Originally named Kiss, Buddy's claim to fame in history is being the first seeing-eye dog in America. Her story really begins in 1908 in Nashville, Tennessee, with the birth of Morris Frank. At the age of six, while horseback riding, he hit a tree branch and lost sight in his right eye. Years later at sixteen, during a boxing match, he lost sight in his left eye, rendering him completely blind. In November 1927, Frank's father discovered an article by Dorothy Eustis, an American philanthropist in Switzerland, about running a school that supplied rescue dogs to the police and the Red Cross, and also describing schools in Germany training guide dogs for World War I veterans blinded from mustard gas. Frustrated with having to depend on other people for help, Frank wrote to Eustis asking for help, not only for himself but also for the possibility of setting up a guide dog training center in America; she agreed. Frank traveled to Switzerland and trained with a female German shepherd, and within a few weeks he was able to navigate the Swiss village. In June 1928, Frank returned with Buddy to New York City, where he demonstrated Buddy's abilities to the media by successfully crossing a busy New York street. In January 1929, Frank and Eustis set up the first guide dog school in the U.S. The Seeing Eye opened

John Seward Johnson's statue of Buddy, the first seeing eye dog in America, with her owner Frank Morris in Morristown, New Jersey (CC BY-SA 4.0).

in Nashville and after two years relocated to New Jersey for better weather for training dogs. Between 1929 and 1956, Frank traveled throughout the U.S. campaigning for equal-access laws for people with guide dogs. By 1956, every state had passed laws allowing blind people with guide dogs access to all public spaces. Buddy died on May 23, 1938, and on April 29, 2005, a sculpture of Morris Frank and Buddy titled *The Way to Independence* was presented in Morristown, New Jersey. Frank named all of his guide dogs Buddy.

Bum (Dog)

In July 1908, a yellow dog named Bum was rescued on the streets of Little Italy by Patrolman Cornelius O'Neil, and he became the mascot of the Twelfth Precinct of the New York City Police Department. Bum served in the police department for more than eight years and is credited with saving many lives. When responding to fires, Bum led help to those trapped. In July 1912, Bum was presented with a bronze medal inscribed, "Bum, Twelfth Precinct" on one side, and the other side featured a female figure with a dog. The medal was awarded by the Bideawee Home, a nonprofit humane group rescuing and providing care for stray and unwanted cats and dogs.

Bummer and Lazarus (Dogs)

In the 1860s, Bummer and Lazarus were two stray dogs who became best friends roaming the streets of San Francisco. Bummer, a short-legged black and white dog, earned his name by wandering Montgomery (the main street for saloons and restaurants) begging for food. Lazarus, a yellow-black dog, was rescued by Bummer while being attacked by a larger dog; Bummer chased the dog off. Afterwards Bummer nursed the injured dog back to health by bringing him food

and curling up next to him at night to keep him warm. The injured dog was given the name Lazarus because he seemed to have risen from the dead like Lazarus in the Gospel of John. The two formed an inseparable friendship. In 1862, an ordinance was passed that all dogs not claimed after a certain period of time be destroyed. However, Bummer and Lazarus were exempted from the animal control ordinance because they earned their keep by being excellent at killing rats, which were a big problem in the city. In 1863, Lazarus died with Bummer by his side after being poisoned with rat-bait laced meat. Two years later, Bummer died after being kicked down a flight of stairs. Both dogs were stuffed and displayed in the saloons that they frequented when they were alive. In 1906 the bodies were placed in storage at the Golden Gate Park Museum and were destroyed in 1910. In 1992, a brass plaque was placed in Transamerica Redwood Park with an inscription reading, "Their devotion to each other endeared them to the citizenry.... They belonged to no one person. They belonged to San Francisco.... Two dogs with but a single bark, two tails that wagged as one."

Bunny (Horse)

A children's book for grades 1 to 3, *Bunny the Brave War Horse* by Elizabeth MacLeod relays the true story of a war horse during World War I named Bunny. Bunny, named for his long ears, was a police horse in Toronto, Ontario, Canada, who, along with eighteen other police horses, was sent to fight in the 9th Battery Canadian Field Artillery. As a police horse, he was ridden by Officer Thomas Dundas, who followed Bunny along with Tom's brother Bud to the battlefield at Ypres, Belgium. It was there that the Germans launched a poison gas attack and Bud was killed. After the war ended on November 11, 1918, Bunny was sold to a farmer in Belgium. The book does speak to some of the realities of war. One battle is described in the book: "The sky was midnight black. Drenching rain and thick smoke from the guns made it almost impossible to see. Bombs exploded all around. The noise was terrifying.... Tom visited Bunny in the stable one night. He noticed that the horse beside Bunny was shivering. The horse had been so hungry it had eaten its blanket."

Burma Queen (Pigeon)

Burma Queen saved hundreds of U.S. soldiers under attack at the border of Thailand and Burma. Allied Forces had destroyed their radios and codes to prevent capture by the Japanese. After a successful search for the group, the pigeon parachuted from a B-25 Mitchell bomber and flew over 320 miles through the Burmese mountains to send messages for help back to Allied headquarters.

Buster (Dog)

Buster was a messenger dog with F Company, 155th Infantry Regiment, on Morotai Island. During two trips, traveling through heavy machine gun and mortar fire, he was able to deliver a message with instructions for the patrol to hold its position. Thus he was responsible for reinforcements leading to the destruction of an enemy force and credited with saving the lives of seventeen men.

Butch *see* Tamworth Two

Buttermilk (Horse)

Buttermilk was a buckskin Quarter Horse owned and ridden by cowgirl star Dale Evans in many Western films. In the 1950s television series *The Roy Rogers Show* Evans rode Buttermilk alongside her husband, Roy Rogers, who rode his Palomino, Trigger. Being extremely popular, both horses inspired a huge marketing success with products made for both children and adults. Buttermilk died in 1972 and his hide was stretched over a plaster statue and displayed at the Roy Rogers–Dale Evans Museum in Victorville, California, which later relocated to Branson, Missouri. In 2010, the Branson museum closed and all exhibits were sold.

Byerley Turk (Horse)

Byerley Turk was one of three Arabian horses from which all modern Thoroughbreds are descended. However, it is not really verified that Byerley Turk was a purebred Arabian because he was obtained during combat. In 1686, at the Siege of Buda, the British defeated the Turks. British Captain Robert Byerley seized the horse from a Turkish officer. After retirement the horse was put out to stud and over time sired many great race horses that may or may not have a pure bloodline.

Cabot (Dog)

Cabot, a bulldog mix, was a famous sentry dog who uncharacteristically fled his post to chase down a German messenger dog across "no-man's land" to retrieve German dispatches.

Caesar (Dog)

Caesar of Notts, a wire fox terrier born in 1898, was sired by Cackler of Notts from the kennels of Kathleen, Duchess of Newcastle. In 1902 he was a gift from Lord Dudley to King Edward VII to replace the king's dog Jack, who died after choking on some food. During his life Caesar was given the royal treatment, having a footman assigned to him, sleeping on an easy chair next to the king's bed, and wearing a collar reading, "I am Caesar. I belong to the King." It seems as if Caesar and the king had a really special bond. Caesar would always greet the king by jumping up and down with excitement, and the king would reply, "Do you like your old master, then?" The king never hit the dog; instead he shook his walking stick at him, calling him a "naughty dog." The king asked Lawrence Wrightson, his personal assistant, to care for Caesar should he die before the dog. The House of Faberge carved hardstone models of the king's favorite dogs and racehorses. Caesar's model included his collar and was made of chalcedony, rubies, enamel and gold. The king died before the finished model was delivered; the model was then purchased by Dame Margaret Greville, who gave it to Queen Alexandra. There was also a painting of the dog created for the king by Reuben Ward Binks. After the king's death on May 6, 1910, Caesar refused to eat, spent time whining outside of the king's bedroom and would sneak into the king's bedroom and lie under his bed. The queen managed to get him to eat again, and he was restored to health. At the disapproval of Kaiser Wilhelm II of the German Empire, Caesar led the funeral procession ahead of heads of state including King George V and eight other kings. After the death of King Edward, Caesar remained in the household until his death in April 1914. Shortly after the king's funeral, an unofficial book titled *Where's Master?* was published; it had been written from the "first-person" perspective of Caesar during the time from the king's death up to the funeral. The book was very successful, leading to nine reprints in the first year after publication. Caesar was so loved that yet another portrait was painted by Maud Earl titled *Silent Sorrow*, depicting his head resting on the king's favorite chair. Also the Steiff company began producing stuffed toys of Caesar. The Faberge figurine is now a part of the Royal Collection. In the sculpture of the king and queen on top of their tomb in St. George's Chapel in Windsor Castle, Caesar lies at the king's feet. Caesar's grave is at Marlborough House.

Caesar (Dog)

Caesar, a Red Cross-trained bulldog, and his handler "Uncle Tom" served during World War I, rescuing wounded soldiers from "no-man's land" at the Battle of the Somme. Caesar's story was told to Patricia Stroud (author of *Caesar the ANZAC Dog*) by her mother, Ida, who was present when the New Zealand Rifle Brigade boarded their troop ship to transport the soldiers to Egypt and the Western Front. Caesar was the troop's mascot, wearing on his collar a blue ribbon given to him by Ida.

Caesar (Dog)

Caesar, a German shepherd, was one of 10,000 dogs serving in the United States military during World War II. Max Glazer's family donated Caesar to the military as part of the Dogs for Defence program. Caesar served in the First Marine Dog Platoon in the Second Marine Raider Regiment, deployed from the fall of 1943 until January 23, 1944, to the Pacific island of Bougainville. Fifty-five men and 24 dogs made up the regiment, and the dogs were known as Devil Dogs. One morning in November 1943, the Japanese launched an attack on the Marines in the jungle of Bougainville. All the dog handlers' dogs were keeping watch in their foxholes. Marine PFC Rufus Mayo was in a panic because his dog Caesar was missing. Caesar had heard the enemy coming, instinctively jumped out of the foxhole and was shot three times; in all the chaos Caesar got lost. In their search, Mayo and another Marine followed a trail of blood leading to Caesar's other handler, PFC John Kleeman. Doctors removed two bullets from Caesar but left one bullet lodged by his heart, and in three weeks he was back on duty. Several times during his service Caesar proved the value of dogs in war. On one deployment Caesar had to run messages back and forth between his handlers when their walkie-talkies were unusable due to heavy rains. Another time Caesar saved Mayo's life from a grenade attack. Afterward the Marine wrote home to his family: "I would not give Caesar up for a general's commission." Before storming the beaches on November 1, 1943, the commanding officer of the Marine Raiders, Lt. Col. Alan Shapley, reportedly said to his men, "I want you men to remember that the dogs are least expendable of all."

Cairo (Dog)

On May 2, 2011, Cairo, a Belgian Malinois, participated in Operation Neptune Spear, accompanying the United States Navy SEALs who killed Osama Bin Laden, leader of Al-Qaeda. The SEALs brought Cairo along to detect bombs, search out concealed enemies, and find secret doors or passageways in the compound. Cairo helped to secure the perimeter around the dwelling. On the night of the raid, Cairo wore a customized body vest made of Kevlar. His equipment also included a night-vision camera with a 180-degree field of vision. President Obama met Cairo four days after the raid when he learned that a dog had been involved. *Time* magazine gave Cairo its prestigious Animal of the Year award in 2011.

Calamity (Dog)

Private Peter Bodette, a soldier in the 28th Wisconsin Volunteer Infantry in Pine Bluff, Arkansas, found a puppy with black on his sides and back and yellow underneath. One day in the summer of 1864, Captain Thomas N. Stevens rode into camp and the dog started barking, spooking the captain's horse. A very angry captain shot at the dog, wounding him with two of the shots. He was nursed and survived. When Colonel Edmund B. Gray heard about the incident, he named the dog Calamity. Calamity helped the soldiers forage for food by chasing down hogs; he would hold the hog by the ear until the men could come and kill it. Soldiers were not allowed to shoot hogs. Calamity was given to Bodette since he had found the dog and took care of him during the war. It is believed that after the war Calamity returned to Wisconsin with Bodette.

Canelo (Dog)

Canelo, a symbol of loyalty in Spain, has a monument in his honor in Cadiz and also a street named after him. Canelo was an inseparable companion to an old man, and the two went everywhere together. One of the places the pair would walk to was the hospital where the man received dialysis treatments; Canelo would wait at the hospital's exit to go home. One day while at the hospital the man died of complications from his dialysis. For the next twelve years, Canelo kept vigil every day at the front of the hospital, waiting for his master.

Caporal (Dog)

Caporal, a guard dog, was responsible for the capture of at least two prisoners as mentioned in *Animal Heroes of the Great War* (1925) by Ernest Harold Baynes. One prisoner was pretending to pick wildflowers during an exercise period, and Caporal bit him in his thigh, which drove the prisoner to rejoin his group. Another time, four prisoners escaped and the guards caught three; Caporal caught the fourth, tracking him by scent. It seems the French initially had objected to the use of guard dogs, but when a few dogs were unofficially used prior to 1918 in a prison in Dijon, they began to employ them.

Captain Lederman (Pigeon)

Like many other pigeons, Captain Lederman was parachuted or dropped by bamboo container to gather information behind enemy lines. With this information attached to his leg, he flew over rugged mountains and jungles delivering vital intelligence to Allied command.

Casper *see* Commuting Animals of England

Castor and Pollux (Elephants)

These two elephants, Castor and Pollux, named for the Gemini twins, were killed and sold by the pound to local butchers. The year 1870 began one of the darkest periods in the history of France, referred to as the Siege of Paris, when German forces surround the capital. With no supplies coming in, the French resorted to butchering dogs, cats, and even zoo animals, including Castor and Pollux. For example, menus included "Epaules et filets de Chien braises" (charred dog shoulders and tenderloins) and "Civet de Chat aux Champignons" (cat stew with mushrooms). A rich English politician and writer, Henry Du Pré Labouchère, wrote after sampling an elephant, "Yesterday, I had a slice of Pollux for dinner. Pollux and his brother Castor are two elephants, which have been killed. It was tough, coarse, and oily and I do not recommend English families to eat elephant as long as they can get beef or mutton."

Catarina (Cat)

Edgar Allan Poe loved cats. His beloved tortoiseshell cat, Catarina, is said to have sat on Poe's shoulder watching him write, and to have slept on his wife Virginia's chest to keep warm, which was the inspiration for Poe's horror story about Pluto, "The Black Cat," in December 1846. Catarina died about the same time that Poe passed away. Most people assumed Catarina was a black cat.

Cecil (Lion)

Cecil was a southern African lion living at the Hwange National Park in Matabeleland North, Zimbabwe. When Cecil was killed by American dentist and recreational big-game hunter Walter J. Palmer from Minnesota, it spurred international media coverage and brought to the surface the issue of trophy hunting in the world. The killing sparked outrage from animal conservationists, condemnation by politicians and celebrities, and hate messages from the public. In June 2015, Palmer paid Theo Bronkhorst, a professional hunter-guide, $50,000 to aid him in killing a lion. After being killed, Cecil was skinned and his head removed. Since Palmer had a permit, he was not charged with any crime. In 1999, scientists had tagged 62 lions, including Cecil, with GPS tracking collars for a study of the Wildlife Conservation Research Unit at the University of Oxford and had followed Cecil's movements since 2008. During the study period, 34 of the 62 lions had died, 24 from sport hunting. In 2013, 49 lion carcasses were exported from Zimbabwe as trophies. As a result of the killing of Cecil, conservation organizations made proposals for bills to ban imports of lion trophies to the U.S. and European Union. American, Delta and United Airlines voluntarily banned the transport of hunting trophies. Globally about 1.2 million people signed an online petition, "Justice for Cecil," calling for the Zimbabwean government to stop issuing hunting permits for endangered species. In Zimbabwe, Cecil's killing had a huge negative impact on tourism, causing many tourists who had planned to see Cecil to cancel their trips. On August 1, 2015, Zimbabwe Minister of Environment Oppah Muchinguri banned the hunting of lions, leopards and elephants and all bow-hunting in areas outside of Hwange National Park. Cecil's death sparked animal conservation measures and discussions among government officials in the

United States and other countries. According to Wayne Pacelle, president of the Humane Society of the United States, Cecil had "changed the atmospherics on the issue of trophy hunting around the world." *The New York Times* wrote, "The killing of Cecil … seemed to galvanize public attention."

Champion (Horse)

Champion initially belonged to actor Tom Mix, but was purchased by singing cowboy Gene Autry after Autry starred with the horse in *The Phantom Empire* series. Besides performing numerous tricks such as jumping through hoops, Champion was a celebrity himself, starring in his own television series, *The Adventures of Champion*. The theme song cements his reputation as "Wonder Horse":

> Like a streak of lightnin' flashin' 'cross the sky,
> Like the swiftest arrow whizzin' from a bow,
> Like a mighty cannonball he seems to fly.
> You'll hear about him ever'where you go.
> The time'll come when everyone will know
> The name of Champion the Wonder Horse!

From the 1930s through the 1950s, Champion (and his successors) appeared in one hundred films and television shows. In 1949, Gene Autry had his handprints and Champion (most likely Champion II) had his hoofprints cemented outside Grauman's Chinese Theatre.

Champion Jr. (Horse)

Champion "Champ" Jr., a Sorrel born in 1942, was Gene Autry's second screen horse, appearing in films until 1950. His onscreen billing was "Wonder Horse of the West" and at Columbia Pictures, "World's Wonder Horse." His most memorable role was in the film *The Strawberry Roan*. Champion Jr.'s handler Jay Barry claimed that the horse had a good disposition, which is a great quality in a horse. Quoting Barry is the best way to describe

Champ Jr.: "He doesn't get excited about anything. You can take him to a hospital full of kids, and as long as he has enough room to move, he never bothers. One time, in a town in Iowa, I left Champ to eat lunch. When I came back, there were seven kids sitting on his back. You'd be in plenty of trouble in a case like that if you had a bad horse." It amazed Barry how Champ Jr. was able to travel sometimes as much as seven hundred miles in a day in only a thirty-five-foot trailer, even though it was air-conditioned. His performances included bowing, saying his prayers, playing dead, dancing, and jumping on top of a baby grand piano. Barry said it was easy to get him to do things that would be difficult for other horses, such as going into a building using steps as opposed to a ramp, and riding in an elevator. Barry said, "But he has that kind of disposition…. It's just as though he says, 'Wherever you go, I'll go too, if I have to walk a plank.'" Champion Jr. died in 1977 and was buried on the Melody Ranch.

Charlene Mooken *see* Cincinnati "Cinci" Freedom

Charley (Dog)

This large, ten-year-old standard poodle was John Steinbeck's traveling companion in 1960 on his 10,000-mile tour of the United States. The journey, described in Steinbeck's *Travels with Charley: In Search of America* (1962), was made in a pickup truck, which Steinbeck named Rocinante after Don Quixote's horse; it was fitted with a small cabin in the back. When traveling through Yellowstone National Park, the usually gentle Charley showed a sign of his canine instincts by barking crazily at the bears he saw. Charley was an eager traveler and loved road trips. While planning and preparing, Steinbeck could tell that Charley was nervous that he might not be going with his owner. Charley had a habit of waking early in the morning and wanted Steinbeck to get up early as well. Sometimes Charley would sit quietly beside the bed and stare into the author's face with a sweet look. After breakfast Charley went back to sleep. Charley died in 1961 and was buried on the grounds of Steinbeck's Pacific Grove cottage.

Charley (Horse)

George Catlin rode Charley, a Comanche wild breed pony, across 500 miles of prairie in the 1830s to study and paint Native Americans. He described getting acquainted with Charley through the Native Americans. Stopping at dusk near the bank of a stream, Catlin tethered Charley to a picket fence, but the pony slipped free. After chasing him for half a mile, Catlin gave up, returned to camp and went to sleep. In the middle of the night, he woke up to a shock: a huge figure was looming over him. The next morning Charley was gone again. Catlin spent some time trying to catch him. He finally packed up his things and the saddle and started walking. Charley raced to overtake Catlin, and after that Charley was closely watched and kept under control.

Chase No Face (Cat)

Chase No Face was a bit of an internet star after an automobile accident left her missing a leg and a face. Through Facebook and YouTube she became a source of encouragement to people who suffer from disabilities and disfigurements. Her owner, veterinary assistant Melissa Smith, forewarned houseguests that they may be upset by Chase's appearance. Smith and her children showed this disfigured cat kindness, which in turn demonstrated to others the value of inner beauty and true love. Melissa Smith said, "She didn't have an owner, and I said I would take her home … to get the extra attention and it's been seven years. She's my cat now."

Checkers (Dog)

This cocker spaniel is the Checkers in the famous "Checkers Speech" by Richard Nixon given on September 23, 1952. Senator and Republican vice-presidential candidate Richard Nixon appeared on television to save his candidacy. In his speech Nixon denied accepting money from a secret slush fund and said the only gift he had accepted was a cocker spaniel from a Texas admirer. Nixon spoke of his humble background. The speech went over well with the public and saved Nixon's place as Dwight D. Eisenhower's running mate. Nixon's six-year-old daughter Tricia named the cocker spaniel Checkers. Nixon said that no matter what the dog was staying and would not be returned. Many were moved to tears by his plea for support. The rest is history.

Every year, September 23 is designated National Dogs in Politics Day, which is also known as "Checkers Day." Checkers was the best-known presidential dog to never have lived in the White House.

Cher Ami (Pigeon)

Cher Ami (French for "dear friend") was a Black Check cock carrier pigeon trained in France by the U.S. Army to deliver messages during World War I. After an amazing flight which saved 194 men by delivering an important message to division headquarters, Cher Ami became known as the hero of the 77th Infantry Division (named "Lost Battalion" by the media). On October 3, 1918, 500 men serving under U.S. Army Captain Charles Whittlesey became trapped behind Germany enemy lines in the Argonne Forest and came under fire from the Germans and also friendly fire from Allied troops. Captain Whittlesey dispatched Cher Ami and two other pigeons with a message: "We are along the road parallel to 276.4. Our own artillery is dropping a barrage directly on us. For heaven's sake, stop it!" The three pigeons were dispatched

Cher Ami, a homing pigeon, delivered messages for the U.S. Army during World War I (111-SC-93245, Records of the Office of the Chief Signal Office, National Archives).

and two were killed by the Germans, but Cher Ami escaped and successfully delivered the message despite being shot along the way. He arrived at the destination with a nearly severed leg and blinded in one eye. The army medics were unable to save his leg but gave him a carved wooden leg. During his service with the U.S. Army Signal Corps in France, Cher Ami delivered 12 messages. He was the mascot of the U.S. Department of Service until he died from his battle wounds on June 13, 1919. A taxidermist mounted his body at the Smithsonian Institution's Museum of Natural History.

Chetak (Horse)

Maharana Pratap was often called the "Rider of the Blue Horse" because Indian folklore says that his horse Chetak's coat was tinted blue. During the Battle of Haldighati in 1576 between Pratap's Mewar Empire and the Mughal Empire, Chetak died after carrying Pratap to safety. After Chetak's death, Pratap built a monument where Chetak had died. In 1997, the Indian government commissioned a bronze statue of Pratap riding Chetak, and the national memorial was opened in the city of Haldighati in 2009.

Chick (Mule)

Corporal Hiram Boone's job in the field was to collect ammunition, food and other supplies from air drops and load them onto mules at Camp Landis in northern Burma during World War II. Chick was Boone's personal riding mount from the beginning of his journey all the way to China. Boone cared for Chick and groomed him every day. The military mules endured long treks across tough terrain. When action ended, the mules were ordered to provide the Chinese army with their services. Many of the mules, including Chick, contracted the disease surra and had to be destroyed.

Chinook (Dog)

New Hampshire's official state dog, the Chinook, is a rare breed of sled dog developed in the early 20th century. The Chinook breed was essentially developed by Arthur Treadwell Walden of Wonalancet, New Hampshire. Walden was an experienced sled dog driver and was the lead driver and trainer on Admiral Richard Byrd's 1929 Antarctic expedition. Walden also founded the New England Sled Dog Club in 1924. Walden's beloved lead dog and stud, born in 1917 and named "Chinook," was a result of crossbreeding husky stock with a mastiff-like male. Chinook's main characteristics included a dropped ear and a broad mastiff head and muzzle. Chinook was bred with Belgian sheepdogs, German shepherds, and Canadian Eskimo dogs. At almost 12 years of age, Chinook went missing on Byrd's expedition and was never found. There were several different speculations as to what befell him; one theory was that he wandered off to die, and another theory was that he had a fatal accident, perhaps falling into a crevasse.

Chips (Dog)

Chips was a member of the K-9 Corps during the Second World War. His family, the Wrens, lived in Pleasantville, New York, and donated Chips to the army; he was trained at Front Royal, Virginia, in 1942. Chips' father was a husky and his mother was a mix of collie and German shepherd. He and his handler, Private John Rowell,

were assigned to the 3rd Infantry Division to serve in North Africa, Sicily, Italy, France and Germany. On July 10, 1943, Chips became a hero. On that day, Chips and Powell stood amongst the sounds of war on the Sicilian shore. They were east of the southern coast of Sicily with the British Eighth Army on their right, Canadian troops in the center, and the American Seventh Army under the command of George S. Patton, Jr., on the left. As the soldiers moved up the beach, they came upon a peasant's hut that seemed empty until the silence was broken by machine-gun fire. Chips raced into the hut and managed to take down one of the Italian soldiers by the throat; then three other soldiers came out with their hands up. Inside the hut, Chips was wounded with a bullet from a revolver; he received first aid and stayed on the front lines. That very night Chips performed another heroic act by warning Private

Chips returning home after serving as a member of the K-9 Corps during World War II (National Archives).

Rowell about additional Italians traveling toward the beach, leading to their capture. On November 19, 1943, Chips was awarded the Silver Star, in the words of Captain Edward G. Paar, for "single-handedly eliminating a dangerous machine-nest and causing the surrender of its crew." The award was later nearly revoked due to a military policy forbidding animals to be awarded medals, but the policy was waived in this instance by Major General Lucian K. Truscott, Jr., commander of the Third Division. Being the first dog hero of World War II, Chips had also been awarded the Service Cross and the Purple Heart, as was reported by newsmen in Italy. The subject of giving awards to animals was debated for three months, and in the end, it was decided that only humans could be awarded medals. For heroic actions by animals, "appropriate citation may be published in unit general orders." Chips was sent home two years later due to battle fatigue. It is noted that Chips could have had eight battle stars for the French Moroccan and Sicilian campaigns; for his service in Italy, France, Germany, and Central Europe; and for standing a twelve-hour guard over President Roosevelt and Winston Churchill during their secret meeting at Casa Blanca.

Chiquita (Cheetah)

Josephine Baker (the "Black Pearl"), a dancer during the Jazz Age, had her pet cheetah accompany her to Paris. During her performances, she often had Chiquita join her onstage.

Cholla (Horse)

Cholla was a mustang-Quarter Horse mix born in Nevada on May 20, 1985, and was named after the cholla cactus. Cholla became known as the "painting" horse for his unique talent of creating paintings. Renee Chambers, a trained ballerina,

bought Cholla when he was five. He used an easel and watercolor tins, showing awareness and intention while painting with abstract strokes. After selecting his brush and color, he put his brush in his mouth and began painting using his tongue and teeth. In 2008, Cholla's owner entered a 2004 painting (*The Big Red Buck*) in a competition in Mogliano Veneto, Italy. Awarded an honorable mention by the jury, the horse's artwork was exhibited with 30 other finalists. The Giudecca 795 contemporary art gallery exhibit in Venice from April 24 to September 15, 2009, featured 30 of Cholla's watercolors painted between 2004 and 2009. Cholla was almost 28 when he died on March 22, 2013.

Cholmondeley (Chimpanzee)

This mature chimpanzee, nicknamed "Chumley," was shipped from Africa to the London Zoo by Gerald M. Durrell, who described him in his book *The Overloaded Ark* (1953) as a "fascinating, mischievous, courtly old man." Chumley smoked cigarettes, made "hoo hoos" when pleased and had perfect table manners, until after the meal when he would throw his empty plate and mug as far as possible. He would share things and grunt happily when his offer of a banana was accepted. Chumley became a television star, but when his teeth went bad, he was moved to a sanatorium. After escaping twice and biting a passenger on a bus, he was shot and killed.

Christian (Lion)

Christian was born in captivity on August 12, 1969, at Harrods department store in London. Australians John Rendall and Anthony "Ace" Bourke purchased Christian for 250 guineas, and, along with their friends Jennifer Mary Taylor and Unity Jones, cared for the cub at their house in London. When Christian was a year old and getting larger, the two men moved him to the basement of their furniture store, Sophistocat. The local vicar permitted them to exercise him at the Moravian church graveyard, and they also took the lion on day trips to the ocean. As Christian was growing and the cost of his care was increasing, Rendall and Bourke realized they could no longer keep him in London. They were steered toward Kenyan conservationist George Adamson for assistance. Adamson and his wife Joy agreed to acclimate Christian into the wild at the Kora National Reserve. Christian was introduced to Boy, an older male lion—who incidentally was in the film *Born Free* and was also featured in the documentary film *The Lions Are Free*—and to a female cub, Katania. The three lions would form the nucleus of a pride (a company of lions). After many setbacks in the pride, Christian was the sole survivor. However within a year, a new pride was established with Christian as the head. In 1971, John and Ace traveled to Kenya for a reunion with Christian and were filmed in the documentary *Christian: The Lion at World's End* (released in the U.S. as *Christian the Lion*). After cautiously approaching the two men, Christian greeted them playfully and warmly. In 1973, Adamson informed Rendall that he had not seen Christian's pride in nine months, but the pride returned to Adamson's compound in Kora the day before Rendall's arrival. Rendall described the reunion:

"We called him and he stood up and started to walk towards us very slowly. Then, as if he had become convinced it was us, he ran towards us, threw himself on to us, knocked us over, knocked George over and hugged us, like he used to, with his paws on our shoulders."

That reunion was the last time anyone saw Christian. In his book *My Pride and Joy* Adamson notes that after 97 days, he stopped counting. In 1971, Christian's owners published his story, *A Lion Called*

Christian. In 2010, a children's book, *Christian, The Hugging Lion*, written by Justin Richardson and Peter Parnell and illustrated by Amy June Bates, was published.

Chunee or Chuny (Elephant)

Chunee, an Indian elephant, was 11 feet tall, weighing 7 tons and valued at 1,000 lira at the time he was brought to London in 1809 or 1810. Chunee's career began when he was exhibited at Covent Garden Theatre, then bought by Stephani Polito for his menagerie at Exeter Exchange on the Strand in London; finally in 1817 Edward Cross bought the menagerie. Chunee was a tame theatrical animal appearing in the plays *Blue Beard* at the Theatre Royal in Covent Garden and the pantomime *Harlequin and Padmanaba* and *The Golden Fish* at the Theatre Royal in Drury Lane. Lord Byron stated after witnessing Chunee's performance at the Exeter Exchange on November 14, 1813, "The elephant took and gave me my money again—took off my hat—opened a door—trunked a whip—and behaved so well, that I wish he was my butler." On February 26, 1826, Chunee suddenly became violent and killed one of his keepers—the elephant's actions were perhaps attributed to a bad toothache aggravated by a rotten tusk. His keeper deemed him too dangerous to keep, so on March 1, 1826, he attempted to feed him poison, but when he refused to eat it, soldiers from Somerset House hit Chunee with 152 musket balls. When he did not die, his keeper ultimately killed him with a sword. In the aftermath of his death, Chunee's carcass was butchered and dissected by doctors and medical students, and his skeleton was sold for 100 lira for exhibition at the Egyptian Hall in Piccadilly and also at the Royal College of Surgeons in Lincoln's Inn Fields. His skin, weighing 1,900 pounds, was sold for 50 lira to a tanner. *The Times* printed letters in protest of not only the barbarity demonstrated in Chunee's death but also the poor living conditions of the other animals at Exeter Exchange. The sad events surrounding the elephant's death inspired a play at Sadler's Wells Theatre titled *Chuneelah*, or *The Death of the Elephant at Exeter 'Change*. In 1829, the Exeter Exchange building was demolished.

Cincinnati (Horse)

This famous horse of the Civil War was a son of Lexington, the fastest four-mile Thoroughbred in the United States and a great sire. Cincinnati was also a grandson of the great Boston, who sired Lexington. Cincinnati was a gift from an anonymous admirer during the war. The horse was large (17 hands or 68 inches tall) and powerful and became General Ulysses S. Grant's favorite horse. Being a shy child, Grant easily bonded with horses and excelled in horsemanship at West Point. Cincinnati is best known for being Grant's horse when the general and General Robert E. Lee met to negotiate the surrender of the Confederacy at Appomattox Court House. Cincinnati died in 1878 on the farm of Admiral Daniel Ammen in Maryland. Almost all statues and drawings show Grant astride Cincinnati, including the Ulysses S. Grant Memorial on the Mall in Washington, D.C.

Cincinnati Freedom (Cow)

This Charolais cow, Cincinnati Freedom (AKA Charlene Mooken), went from

Cincinnati, famous for being General Ulysses S. Grant's favorite horse, shown here with Grant in 1864 (Library of Congress).

being nameless and middle-aged to the-cow-on-the-lam to an overnight folk hero. How did she do this? In February 2002, Cincinnati was headed for the slaughterhouse when she jumped over a six-foot fence and made a run for it, eluding animal control for 11 days. She captivated the entire nation and had everyone rooting for her. Eventually she was tranquilized and sent to the SPCA, and that's when she became an overnight hero and was named Charlene Mooken. Cincinnati's Mayor Charlie Luken was given the task of finding her a good home. Fortunately, Peter Max, a New York-based pop-art icon and environmentalist, donated $18,000 worth of original paintings to the SPCA, a sufficient amount to place Charlene (now renamed Cincinnati "Cinci" Freedom) in a home with other rescued farm animals for the rest of her life. In April 2002, Cinci traveled from Ohio to New York's Farm Sanctuary in the Finger Lakes region. In December 2008, Cinci was diagnosed with spinal cancer and had to be euthanized. Since Cinci's escape there have been other cows earning fugitive status from the slaughterhouses: "Unsinkable Molly B," and a dairy cow from Germany, *Yvonne*, who escaped from a Bavarian farm in 2011, living three months in the forest with a herd of deer.

Clara (Rhinoceros)

One month after her mother was killed by Indian hunters, Clara was adopted by Jan Albert Sichterman, director of the Dutch East India Company in Bengal. During the two years she was with Sichterman, she became very tame. In 1740, she was sold or given to Captain Douwe Mout van der Meer. The captain became Clara's agent and companion. On July 22, 1741, Clara traveled to Rotterdam and was put on exhibition to the public for the first time. From this point on, Clara traveled and was exhibited all over Europe until her death in London in 1758 at the age of 20. During her five months in Paris, she inspired the writing of letters, poems, and songs. Several portraits were painted of her throughout her travels. One very sad result of her captivity is that she rubbed off her horn (or it was cut off), but she did eventually grow a new one. The Natuurhistorisch Museum in Rotterdam held an exhibition on Clara in 1991 and in 2008 created the Clara Memorial to mark the 250th anniversary of her death.

Clever Hans (Horse)

Clever Hans (*der Kluge Hans* in German) was an Orlov Trotter horse owned by math teacher, amateur horse trainer, phrenologist and mystic Wilhelm von Osten. Charles Darwin's publications spurned the public's interest in animal intelligence during the early 20th century. Von Osten traveled through Germany claiming that Hans had been taught to: add, subtract, multiply, divide, work with fractions, tell time, keep track of the calendar, differentiate musical tones, read, spell and understand German. Hans would answer questions, both oral and written, by tapping his foot. The German board of education launched an investigation into von Osten's claims. Thus Carl Stumpf, philosopher and psychologist, formed a panel of 13 people (a veterinarian, a circus manager, a Cavalry officer, school teachers, and the director of the Berlin Zoological Garden) known as the "Hans Commission." In September 1904, they submitted their findings to psychologist Oskar Pfungst that there were no tricks in Hans' performances. In a 1907 investigation Pfungst concluded that the horse was watching the trainer's reactions and did not perform the intellectual tasks. In the research methodology the horse was found to be responding to the involuntary cues in the body language of the trainer, who was able to solve the problems. In a number of trials, Pfungst

Clever Hans, shown here in 1904, was made famous by his owner, Wilhelm von Osten, who claimed that the horse could answer questions by tapping his foot (Karl Krall, Denkende Tiere, Leipzig 1912, Tafel 2).

ruled out fraud by finding that even if von Osten was not the questioner, Pfungst observed that the horse could only answer the questions correctly if the questioner knew the answer and the horse could see the questioner. Hans got 89 percent of the answers correct only if von Osten knew the answers, and if he did not know the correct answers, Hans only got 6 percent correct. Pfungst then plunged into examining the questioner's behavior. He found that the questioner's posture and facial expressions changed to reveal tension when the horse's tapping was getting close to the correct answer and then released when the horse tapped out the correct answer. Pfungst also tested human subjects where he would play the part of the horse. The subjects were asked to stand on his right and to concentrate on a number or simple mathematical problem. Pfungst then tapped out the answer with his right hand, and he often observed "a sudden slight upward jerk

of the head" by the human subject upon reaching the final tap. Pfungst's studies resulted in a term called the "Clever Hans Effect," which proved to be very important in the observer-expectancy effect and later studies in animal cognition, showing that animals can respond to subliminal cues. However the studies show, we should take care in assigning attributes to animals that they do not possess. Clever Hans had a very sad end to his life. In 1914 he was drafted into the army, and in 1916 he was killed and thought to be consumed by his officers. Hans is remembered for teaching us a lesson in anthropomorphism (the attribution of human characteristics or behavior to an animal), which is useful to animal rights activists.

Colo (Gorilla)

Colo, born at the Columbus Zoo in Ohio on December 22, 1956, was the first gorilla born in captivity anywhere in the world. Her father, Baron, was 11 years old and weighed 380 pounds; her mother, Christiana, was 9 years old and weighed 260 pounds. Colo was briefly named Cuddles prior to a contest held to officially name her. Colo's name was derived from the place of her birth, **Colu**mbus, Ohio. Zookeepers did not believe it possible for a gorilla to be born and raised in captivity. Her birth was a surprise; she was found on the floor of the cage weighing three and a quarter pounds and needing CPR and was placed in an incubator. She was rejected at birth by her mother; therefore zookeepers had to hand-raise Colo. They raised her much like a human child by dressing her in clothes and feeding her bottles of formula. Colo lived long enough to become a great-great grandmother of three.

Colonel (Dog)

Colonel was a dog belonging to Charles Kestler, owner of the Merchants' Hotel in Denver. The dog made news with an article published in *The New York Times* on May 1, 1876. It is an amusing story about an ingenious dog. Colonel's favorite game was playing ball, but on this particular afternoon everyone was too busy to play with him. Of course he knew that his ball was located in the drawer of the desk in the hotel office, "keeping place." However, his problem was how to get to the ball. According to the reporter witnessing Colonel's actions, "Going under the desk, he manifestly observed the location of the back of the drawer, and also saw (and reasoned) that by standing on his hind legs he could insert his front paws between the back of the drawer and the back of the desk, and so press the former forward, and thus secure his much-coveted treasure, the ball."

Colonist II (Horse)

Winston Churchill developed his love for horses as a junior officer in the cavalry in the 1890s. He became increasingly interested in horse racing as he aged. In 1949, Churchill bought Colonist II, who won thirteen races netting 12,000 pounds in prize money. Churchill even commissioned the equestrian artist Raoul Millais to paint his beloved horse.

Comanche (Horse)

Captain Myles Keogh's mixed-breed personal horse, Comanche, was a survivor of General George Armstrong Custer's force at the Battle of the Little Big Horn in 1876. The 13-year-old horse was seriously wounded in the neck, lung, and groin, but recovered to live seventeen more years. After his recovery, he was paraded at ceremonies and allowed to wander free the rest of the time, and no one was permitted to ride him. After his death from colic in 1891, his body was stuffed and put on display at the University of Kansas. He was believed to be 29 years old at his death. Other horses survived the battle, and those in better shape were taken as spoils of battle. There was also a yellow bulldog survivor. Most likely, Comanche was born around 1862 on the Great Horse Desert of Texas. He was probably rounded up by cowboys for cattle drives and then sold to the army in 1868. Before the Battle of Little Big Horn, Comanche had survived other injuries caused during skirmishes.

Commando (Pigeon)

During World War II, Commando flew ninety missions delivering crucial intelligence into and out of German-occupied France for the British Armed Forces. The pigeons were carried into the war zones by British paratroopers and released as necessary with messages. Commando was awarded the Dickin Medal in 1942 for completing three notable missions. The messages included the locations of German troops, industrial sites, and injured British

Comanche is best known as a survivor of General George Armstrong Custer's Battle of Little Big Horn in 1876 (photograph by John C.H. Grabill, Library of Congress).

soldiers. After the war, Commando enjoyed a celebrity status and participated in an exhibition of wartime homing pigeons.

Commuting Animals of England

1. Casper (Cat)

Owner Susan Finden, who rescued Casper in 2002, had noticed that sometimes the cat went missing for a while. Then one day she went to take the bus and Casper followed her to the stop. The bus driver told her that the cat was a frequent passenger. This cat joined the queue and quietly boarded and disembarked with other passengers. He usually curled up on the back seat for the trip. Casper was killed by a hit-and-run driver on the road outside his home on January 14, 2010.

2. Dodger (Cat)

Dodger, an elderly ginger cat from Bridport, England, became famous in 2011 when he was found to be riding the local buses. He was seen hopping on and off buses, sitting on passengers' laps, and eating discarded food. He always returned home, especially when the driver knew where he lived. He became ill with an inoperable tumor, and in February 2012 he was euthanized. He was about 15 years old.

3. Macavity (Cat)

In early 2007, drivers of bus number 331 in Walsall noticed an unusual passenger: a beautiful white cat with one blue eye and one green eye and wearing a purple collar. The drivers nicknamed him Macavity, after the cat in the poem "Macavity! The Mystery Cat" from T.S. Eliot's *Old Possum's Book of Practical Cats*—the inspiration for the musical *Cats*. Macavity always boarded at a certain stop and then got off the bus again at the next stop near a fish-and-chips shop. He appeared two or three times a week and no one knew why or to whom he belonged. Then in April

2010, the number 331 bus stopped operating. Nothing more is known of Macavity.

4. Ratty (Dog)

Ratty, a ten-year-old Jack Russell terrier, loved to hop on the Number 10 bus and travel five miles to join the regulars at a North Yorkshire pub. He charmed the bar staff at The Black Bull pub in York, where he was fed sausages while sitting with the locals. When new owners of the pub banned him, he started visiting another nearby pub. Ratty was struck and killed by a speeding hit-and-run driver in 2010.

Congo (Chimpanzee)

Congo was born in 1954 and went down in history as the greatest animal painter. While Congo was living at the London Zoo, Desmond Morris, zoologist, author and surrealist painter, observed Congo's abilities at two years old by giving him a pencil and paper. When Morris drew a shape on one side of a piece of paper, Congo would draw the same shape on the other side. Between two and four years of age, Congo created 400 works of abstract art, which were met with some skeptical reactions. However in the art world, his work was praised by some of the greatest artists in history, including Pablo Picasso and Salvador Dalí. Congo's style has been described as "lyrical abstract impressionism," somewhat similar to Jackson Pollock's works. Salvador Dalí gave the famous quote, "The hand of the chimpanzee is quasi-human. The hand of Jackson Pollock is totally animal." Considered to be a natural artist, Congo understood color balance and scrubbing and could make unique pigments. If stopped before he felt a painting was finished, Congo would throw a fit, and when he considered it finished, he stopped. History recognizes Congo to be a bestselling artist. In fact, on June 20, 2005, three of his paintings, alongside works by Renoir and Warhol, were auctioned off at Bonhams and sold to American collector Howard Hong for $26,000. This talented chimpanzee died in 1964 of tuberculosis at ten years of age.

Copenhagen (Horse)

Copenhagen, of a mixed parentage of Thoroughbred and Arabian, was named for the British victory at the Second Battle of Copenhagen. Following a brief racing career, he was sold to the Duke of Wellington, becoming the duke's favorite horse, which he rode at the Battle of Waterloo. The duke said of Copenhagen, "There may have been many faster horses, no doubt many handsomer, but for bottom and endurance I never saw his fellow." After the battle, Copenhagen participated in parades and ceremonies until his retirement to the duke's Stratfield Saye House, where he lived until his death on February 12, 1836, at the age of 28.

The epitaph engraved on Copenhagen's headstone reads:

Here Lies
COPENHAGEN
The Charger ridden by
THE DUKE OF WELLINGTON
The entire day at the
BATTLE OF WATERLOO
Born 1808. Died 1836. *God's humbler instrument though meaner clay Should share the glory of that glorious day.*

Creamy (Dog)

Below is one account demonstrating the importance of messenger dogs during World War I:

The last time this Division (18th) was in action, I was sent to Brigade Headquarters. After being there for one night, my dog Creamy was taken out during an attack and carried a map of—also a message from the front-line trenches back to Brigade Headquarters. Time taken was 25 minutes, whereas a man took from two-and-a-half to three hours. Under the conditions and heavy shell fire, it was very good, and my officers were highly pleased with it, for the map and message were very important and all other means of communication at the time failed [Richardson, *Watch Dogs: Their Training & Management,* 207].

Creme Puff (Cat)

Creme Puff was a female cat recorded as the oldest cat ever. She died in 2005 at the age of 38 years and three days old.

Crimean Tom (Cat)

Crimean Tom (AKA Sevastopol Tom) was a hero in 1854 to British and French troops during the Crimean War. Lieutenant William Gair of the 6th Dragoon Guards found the cat in the Russian port town of Sevastopol. Tom was known for leading the starving troops to food sources hidden beneath the city by the Russian defenders. After the war Gair brought Tom back to England, where he died on December 31, 1856. After Tom died Gair had him stuffed and presented to the Royal United Services Institute. A stuffed cat purchased in the 1950s at the Portobello Road Market by Lady Faith Compton Mackenzie was determined to be Crimean Tom, but there is no substantial proof that it is the same cat.

Crisp V.C. *see* Red Cock

Crump (Dog)

This little Griffon-mix mascot, a gift to General Sir E. Stuart Wortley, was a source of amusement. Crump would be seen sitting on the front of the saddle when the general rode his horse. Also this comical little dog was often seen walking with a lit cigarette in a holder or a pipe between his teeth.

Curly (Dog)

While in Camp Dennison, Ohio, on April 19, 1861, Mrs. Shellabarger gave a puppy, a brown spaniel named Curly, to Private John H. Crouse of Company A, 11th Ohio Volunteer Infantry. Curly became a new recruit and won the hearts of all in the regiment with his intelligence and good looks. On August 17, 1862, orders were given to get rid of all dogs by throwing them in the river because they had become burdens. When the man assigned to the gruesome task approached the 11th Ohio's Company A, the order was revoked and Curly was saved. On October 28, 1862, Curly was nearly killed by a soldier target practicing when he caught a bullet in his neck. Afterwards Curly was nursed back to health. Curly's finest hour occurred at the Battle of Chickamauga in late September of 1863. Curly stayed on the battlefield, going from one wounded soldier to another giving friendly glances to boost morale, even when tempted away with food by Confederate soldiers. After the battle, as the wounded were removed under a flag of truce, Curly made his escape. Curly was missing but later discovered at another camp and reunited with his regiment. On the return trip home in 1864, Curly fell out of the train and broke his leg. Curly survived and after the war he attended regimental reunions. Eventually, due to his age and health, he was sent to live at the Central National Home for Disabled Volunteer Soldiers in Dayton, Ohio. Curly died at twelve years old as a war hero and veteran. His regiment gave Curly a badge with the inscription, "I am Company A's Dog, Whose Dog Are You?"

Daisy (Dog)

Daisy, a mixed-breed mascot of a Norwegian trawler, was aboard when it was torpedoed by the enemy in World War II. Those of the crew not killed were thrown into the icy sea. Daisy joined them in the water and swam up to each man, licking his face and giving him hope and encouragement. Those men who were finally rescued all praised Daisy, who had given them new hope.

Daisy (Giraffe)

Betty and Jock Leslie-Melville purchased Giraffe Manor, a small hotel near Nairobi, Kenya, in 1974, and since then additional land has been purchased, bringing the total to 115 acres. Shortly after purchasing the manor, the Leslie-Melvilles learned that the only remaining Rothschild giraffes, about 180, were endangered because of the government's purchase of land, and consequently the giraffes were being killed. The manor was already home to three wild bull giraffes; the Leslie-Melvilles adopted Daisy, a two-month-old, eight-foot-tall, and 450-pound giraffe. The baby was lassoed and transported 225 miles to a corral at the manor. Daisy gave big, slobbery kisses and loved milk and carrots. At the end of four months, Daisy was nine feet tall and weighed 700 pounds. Daisy was soon joined by another baby giraffe, Marlon (named after Marlon Brando). The manor, along with others, has run a breeding program to reintroduce the Rothschild giraffe into the wild to expand the gene pool. In 1983, Betty's son Rick Anderson and his wife took over management of the hotel. Guests of the hotel could feed the giraffes at the breakfast table, through the front door, and from their second-story rooms. In March 2009, Mikey and Tanya Carr-Hartley purchased the manor, which is now a part of The Safari Collection group of hotels and lodges.

Dale of Cawsalta (Dog)

Sergeant John Cawsey of the Royal Canadian Mounted Police owned and trained Dale, a German Shepherd, to help him in his police work in northwestern Canada. Dale won international fame by tracking a wanted man hidden in a wheat field, and finding a lost child after more than a hundred men had given up—both feats within 24 hours. Because of Dale, the Canadian police officials accepted dogs into the force. In 1935, Dale became the first canine member of the Mounties and was officially known as K470. During Dale's long career, he aided in bringing to justice the murderers of two Mounties and led his master to safety through a blizzard. His most remarkable case was tracking an arsonist during a dust storm from the scent off a piece of underwear and a cotton swab. Dale was made a member of the "Legion of Honor of the Dog World of the United States" by a Chicago magazine and received a citation from the Humane Society.

Dan Patch (Horse)

In 1895, Daniel Messner, Jr., a wealthy dry-goods merchant, bred his lame three-year-old filly Zelica with Joe Patchen, a top-quality racehorse. Their offspring, Dan Patch, a mahogany bay Standardbred stallion, was foaled on April 29, 1896, in Oxford, Indiana. As a colt Dan Patch had a somewhat shaky start in life. He was born with crooked legs and could not stand on his own, and some suggested that he be put down. However, eventually his legs grew straight, but Messner still thought he showed little promise and thought his future was to haul a delivery wagon. Dan Patch may very well have had a dismal future if not for John Wattles, a close friend of Messner, who saw potential and began training the horse and developing his abilities. A fully grown Dan Patch stood 64 inches tall and weighed 1,165 pounds. Dan Patch became a celebrity in harness racing, which was one of the largest sports in the nation at that time. Harness races consisted of multiple heats in which the horse had to win the majority of heats, usually three out of five, to win the race. Dan Patch began his racing career in 1900 as a four-year-old and never lost a race and only lost two heats. In 1900 he entered races in Indiana county fairs; in 1901 Messner entered Dan Patch on the Grand Circuit, and on July 17 he had his first race at the Grosse Point Racetrack in Detroit. At the age of nine, Dan Patch set his official record of 1 minute, 55 and a quarter seconds (1:55¼), which stood for over 30 years. Dan Patch had several owners during his career: In February 1902, Manley E. Sturges, a casino owner in New York, bought him for $20,000; in December 1902, Sturges sold the horse to Marion Willis Savage of Minnesota for $60,000 (equivalent to over two million dollars in 2022). At the end of his racing career Dan Patch traveled in exhibitions and earned millions of dollars in product endorsements. On July 11, 1916, at 10:00 a.m. Dan Patch died of a massive heart attack. Over the years Dan Patch had many memorials; in 1953 he was inducted into the Harness Racing Museum and Hall of Fame. Two noted cultural references include the 1949 movie *The Great Dan Patch* starring Dennis O'Keefe and Gail Russell and the Disney movie *So Dear to My Heart*, which begins with Dan Patch's train pulling into a small town.

Daniel Webster (Horse)

Daniel Webster (nickname "that devil Dan"), a dark bay purebred, was Union General George B. McClellan's favorite war horse. Because Daniel Webster was such a speedy horse and the general's staff had a hard time keeping up with him, they began to call him "that devil Dan." Daniel Webster was with McClellan at Antietam (also known as the Battle of Sharpsburg). The battle was fought on September 17, 1862, between Confederate General Robert E. Lee's army of northern Virginia and McClellan's army of the Potomac. This was the bloodiest day in United States history, with 22,717 soldiers dead, wounded

or missing. Daniel Webster became the McClellan family horse at Orange, New Jersey, after McClellan retired from military service. The horse died at the age of 23.

Dash (Dog)

From 1856 to 1861, Dash was the fire dog of the Good Will Fire Company in Philadelphia. In 1861, Dash became part of the 23rd Pennsylvania Infantry that came to be known as Birney's Zouaves. The regiment spent time training at Camp Graham. During the time in camp, Dash may have been the only Civil War canine to have seen President Abraham Lincoln in person—on two different occasions. Dash gained quite a bit of weight and couldn't keep up on marches. However, the men loved him and shared their rations and took turns carrying him. When under fire Dash stepped up to the turmoil. In 1862, Dash was captured with some of the men after a battle. Later a few men and Dash escaped. He was wounded and considered a brave hero. He was to be sent home but was unfortunately lost and never seen again.

David Graybeard (Chimpanzee)

David Graybeard, easily recognizable by his silver facial hair, was the first to accept the presence of Jane Goodall when she began her ten-year study of wild chimpanzees in 1960 in Tanzania's Gombe Stream National Park. David disproved the belief that only humans are able to make and use tools. David's lack of fear allowed Goodall to observe him eating meat and using, even making, tools. Jane observed him extracting termites from a termite hill using a grass stalk and eating the termites hanging onto it. On another occasion she observed David and another chimpanzee stripping leaves off twigs to make fishing tools. David was the first chimpanzee to visit her camp, take a banana from her hand and let her touch him. Once David showed that he was comfortable with Jane, other chimps warmed up to her. In 2011 Jane Goodall said, "[David] was the first chimpanzee who let me come close, who lost his fear," and "he helped introduce me to this magic world out in the forest." Jane believes that David died during a pneumonia epidemic in 1968. *Time* magazine named David one of the 15 most influential animals that ever lived.

Delta (Dog)

Pompeii, an ancient Roman city near Naples, was destroyed and buried after the volcano Mount Vesuvius erupted in 79 CE. A cast of Delta's body was found lying over the body of a young child. It is believed that she tried to save the child's life following the eruption. Nearby there was an intact silver collar engraved with the name Delta. The child may have been a relative of the dog's owner, Severinus. It is also believed that Delta had saved her master three times. The first time she pulled him out of the sea, saving him from drowning; the second time involved fighting off four thieves; and finally Delta fought off a wolf and chased it away. Assumed to be true but...?

Dewey Readmore Books (Cat)

Dewey was the resident cat of the Spencer Public Library in Iowa. On January 18, 1988, librarian Vicki Myron discovered an abandoned eight-week-old male kitten in the library's drop box. The kitten was adopted by the library, and the staff named him Dewey after Melvil Dewey, inventor of the Dewey Decimal library classification system. "Readmore Books" was added to Dewey's name—inspired by Cap'n O.G. Readmore, the cartoon library cat. The library staff gave him the title of "staff supervisor" and celebrated his birthday on November 18. Dewey's story soon gained fame both nationally and internationally: he was featured in Gary Roma's documentary *Puss in Books* in 1997; on Paul Harvey's radio program *The Rest of the Story* in June 1999; and in a Japanese documentary about cats in 2004. In 1990, Dewey won 80 percent of votes in a charity pet photograph contest. He also appeared in calendars, and by 2005 he had raised $4,000 for the library from sales of postcards sold by the Friends of the Library. Dewey had several health issues including megacolon, arthritis, and hyperthyroidism, but it was the diagnosis of a stomach tumor that led Myron to have him euthanized on November 29, 2006 (at an age equivalent to the nineties in humans). In December 2006, his cremated remains were buried outside of the library after a public memorial service. Vicki Myron wrote several books about Dewey: In 2008, she wrote, with Bret Witter, *Dewey: The Small-Town Library Cat Who Touched the World*, reaching number one on *The New York Times* Best Seller list of nonfiction books. The book tells Dewey's story at the library and how the cat eased the burdens of the town's difficulties and in Myron's personal life. Based on this book, Myron and Witter wrote two children's books: a picture book, *Dewey: There's a Cat in the Library!*, and for middle-grades, *Dewey the Library Cat*. In 2010, Myron and Witter wrote two more books: *Dewey's Nine Lives*, detailing unreleased stories about Dewey and tales about other cats after Dewey's death and targeted for 3–6-year-olds; and *Dewey's Christmas at the Library*. In 2008, Meryl Streep was cast to star in a film based on the book; however, the project was never developed.

Diamond (Dog)

Sir Isaac Newton's beloved dog Diamond took her place in history for delaying Newton's work. In his historical research, respected professor Stanley Coren pieced together the events that led to the destruction of Newton's work. One time Newton went to see who was knocking at his door and closed the door to his study, leaving Diamond alone inside. Hearing an unfamiliar voice, an excited Diamond began running around the room; she hit the desk, causing a lit candle to fall and set fire to an unfinished manuscript. Coren wrote in his book *The Pawprints of History: Dogs and the Course of Human Events*, "It would be close to a full year before Newton would reconstruct the theory of gravity in full, thus an entire year of intellectual life and research, by one of the greatest scientific minds of his era, was lost due to the actions of a dog." The incident, however, did not decrease Newton's love for his beloved dog. He lifted Diamond up in his arms, saying, "O Diamond, Diamond, thou little knowest the damage thou hast done."

Dick (Dog)

Dick was a black retriever-mix messenger war dog. His story illustrates the deep devotion of these messenger dogs. Dick and his handler, Sgt. Herman H. Boude, patrolled 48 out of 53 days, and on hardly any of those days did Dick fail to alert the presence of Japanese soldiers. Despite being badly wounded, he managed to deliver his message and afterwards was treated and returned to duty. Unfortunately, he was not so lucky to escape permanent injury when returned to the front. He became lame, had a bullet lodged between his shoulder and the wall of his chest, and also had a shell splinter in the small of his back.

Dick (Mockingbird)

In the early 1770s, Thomas Jefferson purchased a mockingbird from one of his father-in-law's slaves for five shillings. From then on Jefferson kept mockingbirds for pets. While president from 1801 to 1809,

he had a favorite bird named Dick, who was "the constant companion of his solitary and studious hours," according to Jefferson's friend Margaret Bayard Smith. According to Smith, the bird would eat from the president's lips, and often when Jefferson retired to his chamber, it would hop up the stairs after him, and while he napped, it would sit on his couch and sing. In addition, when Jefferson took out his violin and began to play, Dick would "pour out his song along with the violin." Jefferson is thought to be the first president to have a pet living in the White House.

Digger (Dog)

Digger's story begins when a 16-year-old Matthew left home in Port Melbourne, Australia, for service in World War I. When this young Australian soldier made a quick decision to smuggle his brown and white bulldog, Digger, aboard a troop ship, he had no idea how big an impact the dog would have on his war experience. Matthew worked as a stretcher-bearer at the front, and Digger helped to rescue the wounded. There is a memorial commemorating Digger's service with the Australian Imperial Force during World War I.

Digit (Gorilla)

American zoologist Dian Fossey first saw Digit when he was just a few years old. While living in the mountains of Rwanda with gorillas, she ended up forming a special bond with him, giving him the name Digit when she noticed that one of his fingers was injured. Digit was always playful and curious even as he grew into a large silverback. When Digit was brutally killed by poachers in 1977, Fossey's grief was extreme. In 1978, she founded the Digit Fund in his honor. After Fossey was murdered in 1985, her grave was placed next to Digit's in the forest. Digit sacrificed himself trying to defend his group from the poachers. The Digit Fund is now known as the Dian Fossey Gorilla Fund International and works to fight gorilla poaching. Authorities were never able to determine the motive behind Fossey's killing, but many believe that her death was linked to her anti-poaching activism.

Dixie *see* **Tabby**

Dodger *see* **Commuting Animals of England**

Dolly (Sheep)

Born on July 5, 1996, Dolly became the first mammal to be cloned from an adult cell, which proved that a technique called somatic cell nuclear transfer is possible. This technique involves the transferring of the cell nucleus from an adult cell into an unfertilized egg by administering a blast of electricity and then implanting the egg into a surrogate. At age six Dolly died of a lung disease. She has been called the world's most famous sheep. The cloning technique has since been applied on other large mammals such as pigs, deer, horses and bulls.

Don (Dog)

"Don the Talking Dog" was raised in Theerhutte, Germany, by Martin Ebers and Martha Haberland. Although his breed was unknown, he was thought to have been a German hunting dog, possibly a setter or pointer. By the early 20th century Don had become a popular vaudeville act, able to vocalize eight German words including *haben* (have), *kuchen* (cake), *hunger*, his name, *ja* (yes), *nein* (no), *ruhe* (quiet or rest), and the name of his owner and his fiancée. Several publications wrote about the phenomenal Don the Talking Dog. The December 11, 1910, edition of *The New York Times* described Don as an "uncommonly intelligent animal." In

April 1911, *The Salt Lake Tribune* reported that after examination, "a number of the most eminent zoologists" had declared him "a genuine prodigy." In May 1912, *Science* journal came to the conclusion that Don's "speaking" was more "the production of sounds which produce illusions in the hearer." Being a sensation in Germany, Don was brought to the United States by vaudeville promoter Willie Hammerstein, financing the venture with a $50,000 ($1.2 million today) bond in case the dog died on his way to New York. "This makes Don the most valuable dog in the world," wrote *The New York Times*. In July 1912, Don and his owners traveled to the United States on board the SS *Kronprinz Wilhelm*. For the next two years Don first performed at Hammerstein's Roof Garden Theater in New York City, where he shared the bill with Harry Houdini, and then he toured the country. Don also became a celebrity endorser for Milk-Bone dog biscuits. After two years Don retired to his home near Dresden, Germany, and died in late 1915 at about 12 years old.

Douglas *see* **Old Douglas**

Duchess (Dog)

Duchess (Brand Number 7H74) was responsible for the deaths of 42 Japanese soldiers. She was a member of the 39th Infantry Scout Dog Platoon handled by Sergeant Knight. While on patrol with the 3rd Battalion, 123rd Infantry on April 30, 1945, she was used to inspect an enemy cave on the island of Luzon in the Philippines. At the entrance Duchess sent a strong alert of the presence of the enemy; grenades were thrown into the cave, killing 33 Japanese. On another occasion, while on patrol with the same unit, again Duchess alerted the patrol of some Filipino huts 800 yards ahead. Upon confirmation the patrol used mortar and machine guns, resulting in the deaths of nine Japanese.

Duffy *see* **Murphy**

Dzok (Dog)

Dzok (pronounced "Jock"), a black-haired mixed breed, is famous for being perhaps the most loyal dog in Poland. While driving, his owner suffered a fatal heart attack, and Dzok waited for one year—from 1990 to 1991—at the same spot he had been left for his owner's return. Eventually an old woman adopted him and gave him food, and he went to live in her house. At last he had another caring and loving home, but it was not to last; the woman died and Dzok was taken to a dog shed. After escaping from the dog shed, he was struck and killed by a train. Local sculptor Bronislaw Chromy erected a monument to Dzok in 2001 in Krakow near Wawel Castle.

El Tango (Dog)

El Tango served in the French army. Along with Helda (an Alsatian sheepdog), a famous sentry dog, El Tango followed and exposed kennel assistant Private Vachet of being a traitor for the German army.

Elm Farm Ollie (Cow)

Elm Farm Ollie (known as "Nellie Jay" and post-flight as "Sky Queen"), a Guernsey cow, was born in Bismarck, Missouri, but her fame lives on in Wisconsin. On February 18, 1930, Elm Farm Ollie became the first cow to fly and be milked in an airplane, for the International Air Exposition in St. Louis, Missouri. The flight, in a Ford Tri-Motor airplane piloted by Claude M. Sterling, covered 72 miles from Bismarck to St. Louis. Wisconsin native Elsworth W. Bunce became the first man to milk a cow in flight, with Ollie producing six gallons of milk. A St. Louis newspaper wrote that the cow's mission was "to blaze a trail for the transportation of livestock by air."

Elsa (Lion)

Elsa can be credited with introducing wildlife conservation to the public. George Adamson, a game warden in Kenya, and his wife Joy adopted her as a cub and named her Elsa. George Adamson had shot Elsa's mother on February 1, 1956, thinking she was charging him when in reality she was protecting her cubs. Two of the cubs were sent to the Netherlands' Rotterdam Zoo. The Adamsons raised Elsa at their home in Africa and taught her to hunt and develop survival skills to live in the wild. Elsa survived after the Adamsons released her into the wild, but never really lost contact with her adoptive parents. When she was three years old, she and her three cubs showed up on the Adamsons' doorstep. The Adamsons named the cubs Jespah (male), Gopa (male) and Little Elsa (female). Elsa died two years later from a tic-born disease, babesiosis, and was buried in Meru National Park in Kenya. In 1960 Joy wrote *Born Free*, a nonfiction book about the experience of raising Elsa, and six years later a movie was released promoting wildlife conservation.

Books, films, and television based on Elsa's life:

Books
  Born Free (1960)—Written by Joy Adamson
  Living Free (1961)—Written by Joy Adamson
  Forever Free (1962)—Written by Joy Adamson
  Bwana Game (UK title, 1968); *A Lifetime with Lions* (USA title, 1970)—Written by George Adamson
  My Pride and Joy (1986)—Written by George Adamson

Films
  Elsa the Lioness (1961), 29 minutes; BBC

documentary produced, hosted and
narrated by David Attenborough

Born Free (1966), 95 minutes; George
Adamson was the technical advisor

Living Free (1972), based not on the
book by the same name, but on the
third book of the series, *Forever
Free*

Elsa and Her Cubs, 25 minutes;
extremely rare film footage of Elsa
and her cubs Jespah, Gopa, and Little
Elsa

Elsa's Legacy: The Born Free Story
(2010), 53 minutes; documentary
marking the 50th anniversary of
Joy Adamson's book *Born Free*. It
includes home footage of Elsa and
her cubs shot by the Adamsons and
interviews with Virginia McKenna
and David Attenborough.

Television

Born Free (1974), television series based
on the film, starring Gary Collins
and Diana Muldaur

*Elsa: The Lioness that Changed the
World* (2011), documentary narrated
by Richard Armitage

Elsie (Cow)

Elsie (AKA "You'll Do Lobelia"), a
Jersey heifer, was born in 1932 at Elm Hill
Farm in Massachusetts. She is best known
as the "jubilant, daisy-necklace-wearing"
face of Borden. At the New York World's
Fair in 1939, Borden was exhibiting its
dairy machinery, including the rotolac-
tor; however fairgoers were mostly interested in the true identity of Elsie. Out of
the 150 cows taking part in the display, the
Borden representatives, under pressure,
selected the most attractive and alert cow
and Elsie was launched as the Borden mascot. After the fair, she made public appearances around the country. In 1940, Elsie
appeared in the film *Little Men*, and the
same year she "married" Elmer of Elmer's
Glue fame and gave birth to a calf named

Beulah. In 1941 Elsie suffered serious injuries to her spine in a traffic accident and
had to be euthanized. After a period of
time Elsie was replaced by Borden, and
her successor gave birth to Beauregard
inside Macy's Manhattan flagship store.
Engraved on Elsie's gravestone are these
words: "You'll Do Lobelia-A Pure Bred Jersey Cow, One of the Great Elsies of Our
Time, 1932–1941."

Emily (Cat)

Emily, "The Wandering Cat," definitely lived up to her reputation when she
vanished from her Wisconsin home in
September 2005 and showed up in Europe.
It is a mystery what really happened, but
it seems that while roaming her neighborhood she explored a container holding
paper bales, which was sent to Chicago and
then shipped to Belgium. Emily eventually
found herself at a laminating company in
Nancy, France. Identified by her tags, she
was flown home in business class.

Endal (Dog)

Endal, a pedigree yellow Labrador
retriever, went down in history as one of
the greatest assistance dogs in the world.
Among his other special honors, he
became known as the dog of the millennium and was awarded the PDSA's Gold
Medal "for animal gallantry or devotion
to duty," the peacetime equivalent of the
Dickin Medal. Endal received these honors due to his abilities as an operational
assistance dog. He understood over a hundred instructions, could go shopping,
operate electrical switches, run a washing
machine, and even use an ATM machine.
In the 1990s, Endal was assigned to disabled ex-Royal Navy Chief Petty Officer
Allen Parton, who sustained serious head
injuries, including a loss of 50 percent of
his memory, serving in the Gulf War.
Endal helped Parton with daily activities

and even saved his life. A passing car caused Parton to fall out of his wheelchair, leaving him unconscious. Endal pulled him to safety, covered him with a blanket, and retrieved his mobile phone, barking as he ran to a hotel for help. During his lifetime, Endal was very famous and was featured in both television shows and documentaries around the world. Endal died on March 13, 2009, at the age of 13.

Enos (Chimpanzee)

Enos was the first American to orbit the Earth. On November 29, 1961, he was launched into space aboard *Mercury Atlas 5*; it took one hour, twenty-eight and a half minutes to orbit the Earth. During the flight Enos suffered repeated electric shocks due to equipment malfunction, but did land safely.

Enos was the first American to orbit Earth in 1961 (NASA).

Faith (Cat)

Faith, dubbed "the London Church Cat," is central to an uplifting story during the German Blitz on London in World War II. In 1936, Faith, a stray tabby cat, found her way to St. Augustine's Church while searching for food and warmth. When the verger, Thomas Evans, found her he put her outside. He did this three times, until finally the cat managed to sneak into the church and spend the night. The next morning she showed up in the rectory, and Evans wanted to put her out yet again; however, Father Henry Ross noticed how thin she was and gave her some cream. When no one claimed her, Ross decided she would become his "church cat" and named her Faith; for the next four years she became a permanent fixture in the church and the parishioners loved her. In August 1940, Faith gave birth to a kitten, which was white with black ears and tail. He was named Panda after Chi-Chi, the London Zoo's panda, because his coloring was similar. On September 6, Faith began her quest to settle in the basement. After three times returning the kitten upstairs and Faith bringing him back to the basement, Rosalind, the verger's wife, speculated that Faith must have felt that Panda was in danger. The following night, September 7, a heavy air raid befell the city, destroying homes and killing more than 400 people. On Monday, September 9, Henry went to Westminster on business, and upon his return that evening the air-raid sirens sounded, and Henry spent that night in a shelter. The next morning he returned to find the church in wreckage, with flames rising from the timbers. The firemen told him the cats were most likely dead and asked him to leave the area. However, not giving up so easily, Henry snuck in, climbed over debris and heard a faint meow; with tears of relief he pulled the unharmed cats to safety. Soon after, the roof collapsed, leaving a pile of debris in the corner where the cats had been. By November 1, the church was able to resume limited services. When full grown, Panda became a loved pet at a residential nursing home in Herne Hill. In 1945, Rosalind felt that Faith deserved a medal for her bravery, and she wrote to the People's Dispensary for Sick Animals. Mrs. Dickin replied that a special silver medal would be made and she would present it herself.

On October 12, 1945, Maria Dickin and the archbishop of Canterbury placed the medal around Faith's neck and read the citation:

"From the P.D.S.A. to Faith of St. Augustine's, Watling Street, E.C.

For steadfast courage in the Battle of London, September 9, 1940."

Faith also received another silver medal from the Greenwich Village Humane League of New York. On September 28, 1948, while lying by the fire, Faith quietly passed away at the age of 14. The next morning at 10:30 a.m. there was a short service; Henry blessed the box holding Faith's body while burying it near the churchyard gate.

In tribute to Faith, Father Ross had her photograph framed and hung on the chapel wall, with this text underneath:

Faith

Our dear little church cat of St. Augustine and St. Faith.

The bravest cat in the world.

On Monday, September 9th, 1940, she endured horrors and perils beyond the power of words to tell.

Shielding her kitten in a sort of recess in the house (a spot she selected three days before the tragedy occurred), she sat the whole frightful night of bombing and fire, guarding her little kitten.

The roofs and masonry exploded. The whole house blazed. Four floors fell through in front of her. Fire and water and ruin all around her.

Yet she stayed calm and steadfast and waited for help. We rescued her in the early morning while the place was still burning, and

By the mercy of Almighty God, she and her kitten were not only saved, but unhurt.

God be praised and thanked for His goodness and mercy to our dear little pet.

Note: Despite extensive research, no reference was found regarding the ceremony when Faith was presented with her PDSA medal. However, the dates of the church bombing and Faith's death are correct and were reported in *The Times*. It is unknown what became of Faith's medals and citation.

Fala (Dog)

In 1940, President Franklin D. Roosevelt was given a Scottish terrier puppy named Big Boy. Roosevelt renamed him Murray the Outlaw of Falahill in honor of a distant Scottish ancestor. "Fala" slept in the president's bedroom and often traveled with him. Fala was present when Roosevelt conferred with Winston Churchill aboard the USS *Augusta* in 1941. Fala also became an honorary private of the U.S. Army by contributing $1 to the war effort every day. During the Battle of the Bulge, American soldiers used Fala as a supplementary safeguard against German soldiers. Republicans spread a false rumor that Fala was accidentally left behind on the Aleutian Islands and

President Franklin D. Roosevelt with his Scottish terrier, Fala, August 8, 1940 (Franklin D. Roosevelt Presidential Library & Museum).

that Roosevelt had sent a Navy destroyer to retrieve him. Roosevelt addressed the rumor in a speech in 1944, stating that Fala "resented" those attacks and that the dog's "Scotch soul" was furious at the falsehood. The speech became known as the "Fala speech." In the minutes after Roosevelt died in 1945, Fala ran directly to the front screen door, broke it and ran barking up into the hills; he could be seen standing alone and unmoving. Suffering from deafness and failing health, he lived with Eleanor Roosevelt until his death in 1952. He was always waiting for his master to return. A statue of Fala is featured in the Franklin Delano Roosevelt Memorial in Washington, D.C. In fact there are a total of three statues of Fala. There is also an MGM movie about a typical day in the White House featuring Fala.

Fancy *see* **Little Sorrel**

Fanfreluche (Horse)

Fanfreluche was a bay mare born in Canada on April 9, 1967. She was named by her French-Canadian owner, Jean-Louis Levesque, for the title character of a popular children's television show on the French-language network of the Canadian Broadcasting Corporation. She was a successful race horse in Canada and was sold to an American horseman. She was bred numerous times. In June 1977, while a foal to Secretariat, she was kidnapped from Paris, Kentucky. An extensive search was instigated by the FBI. It took six months but finally she was found 158 miles south, not far from the Tennessee border, living with a family who said that they had found her wandering along a country road. The family had renamed her Brandy and were using her as a riding horse. She was safely returned and in the spring of 1978 she gave birth to a colt named Sain Et Sauf (in English it means "safe and sound"). Fanfreluche died in

July 1999 of old age and was buried at Big Sink Farm in Midway, Kentucky. It took years before the man who kidnapped her was caught. He was paid to kidnap her but never said by whom. He let her go when the pressure got to be intense. Fanfreluche had six major wins as a racehorse, earning $238,688. Her awards include: TRA United States Champion 3-year-old Filly (1970), Canadian Horse of the Year (1970), and Sovereign Award for Outstanding Broodmare (1978). Her honors also include the Canadian Horse Racing Hall of Fame (1981).

Faro (Dog)

Detective dogs Faro the Beauceron and Max the Doberman worked with the French Seventh Army in 1916 to capture the outlaws who terrorized the Department of the Upper Saone. The army was made up of infantry soldiers, engineers, explosives and poisonous gas specialists, MPs, police commissaries, secret service and the head of the intelligence service. Both dogs were experienced sentry workers. Faro was sent to search the abandoned buildings in the woods to no avail. Max followed a trail to an abandoned farmhouse, where a man and woman were living. After interrogation, the man confessed and the gang was caught.

Félicette (Cat)

On October 18, 1963, at 8:09 a.m., Félicette—nicknamed "Astrocat"—a black and white stray cat, was the first and only cat to be launched into space and *survive* the spaceflight. On that day the Centre National D'Études Spatiales was actually set to launch another cat, named Felix; however he somehow went missing and Félicette stepped in. It was a suborbital flight lasting 13 minutes, reaching a height of nearly 98 miles and including five minutes of weightlessness. A pet dealer had found Félicette on the streets of Paris, and then she was purchased by the French government. Félicette joined 13 other cats in training for space-flight by the Centre d'Enseignement et de Recherches de Médecine Aéronautique (CERMA). While the cats trained in high-G centrifuges and compression chambers, they all had permanent electrodes implanted in their brains to study neurological activity. Three months after her successful flight, Félicette was euthanized so scientists could study her brain. Félicette's legacy was to bring France into the space race, yet she was only commemorated on postage stamps. In 2017, Matthew Serge Guy began a campaign to fund a bronze statue of Félicette to note her contribution to science. Sculptor Gill Parker's statue of a cat atop a sphere of the Earth and looking at the stars was unveiled in 2019.

Fend L'Air (Dog)

Fend L'Air was a famous black and white setter owned by a sergeant in the French army. On his owner's ship, they landed at Marseilles, and then crossed France into Belgium. The dog was faithful to his owner throughout the battles, and one night a shell exploded and buried the sergeant. Fend L'Air found his owner unconscious, dug his head out from the dirt and rocks, and barked nonstop for three days until they were rescued. He followed the ambulance, exhausted with bloody paws. When the head nurse at the hospital heard the story, she set up a cot for Fend L'Air next to the sergeant. His beloved dog was the first thing the sergeant saw when he regained consciousness.

Ferdie (Phalanger)

Ferdie, a pygmy flying phalanger, was from Bathurst Island and was enlisted

Ferdie became the mascot of the RAAF Spitfire squadron in 1945 (photograph by John Thomas Harrison, June 14, 1945, Netherlands East Indies, Australian War Memorial, accession number OG2824).

in active service with an RAAF Spitfire squadron by his owner, Flying Officer Robert Addison of Elwood, Victoria, Australia. Ferdie eventually became reformed after having a wild youth. At his larrikin, or "rowdy," stage he developed a taste for beer, but once, after twenty minutes of on-and-off drinking, Ferdie fell into a full glass of beer. After that he never took another drink of beer. Ferdie had competition with fifteen dogs, a cat, another possum and a rooster, but he won the role of squadron mascot.

Fido (Dog)

This yellow mixed-breed dog was adopted by the Abraham Lincoln family in 1855. When Lincoln moved to Washington, D.C., in 1861, he left Fido to the family of John Roll. Fido had accompanied Lincoln when he went to town, sometimes carrying parcels or waiting while Lincoln went for a hair trim. Fido was terrified of fireworks, bells, and crowds, which is why he was left behind with the Rolls. They promised to return Fido when the Lincolns came home from Washington. Fido's favorite sofa was left behind so he would feel at home, and he was permitted to join the Roll family at mealtimes. Lincoln periodically received reports on Fido's new life. After Lincoln was assassinated, Fido was there to say goodbye to his master. A year after Lincoln's assassination, Fido was stabbed to death by a drunken man. There is some discussion as to why the man stabbed Fido.

Fido (Dog)

In November 1941, while on his way home from a bus stop in Borgo San

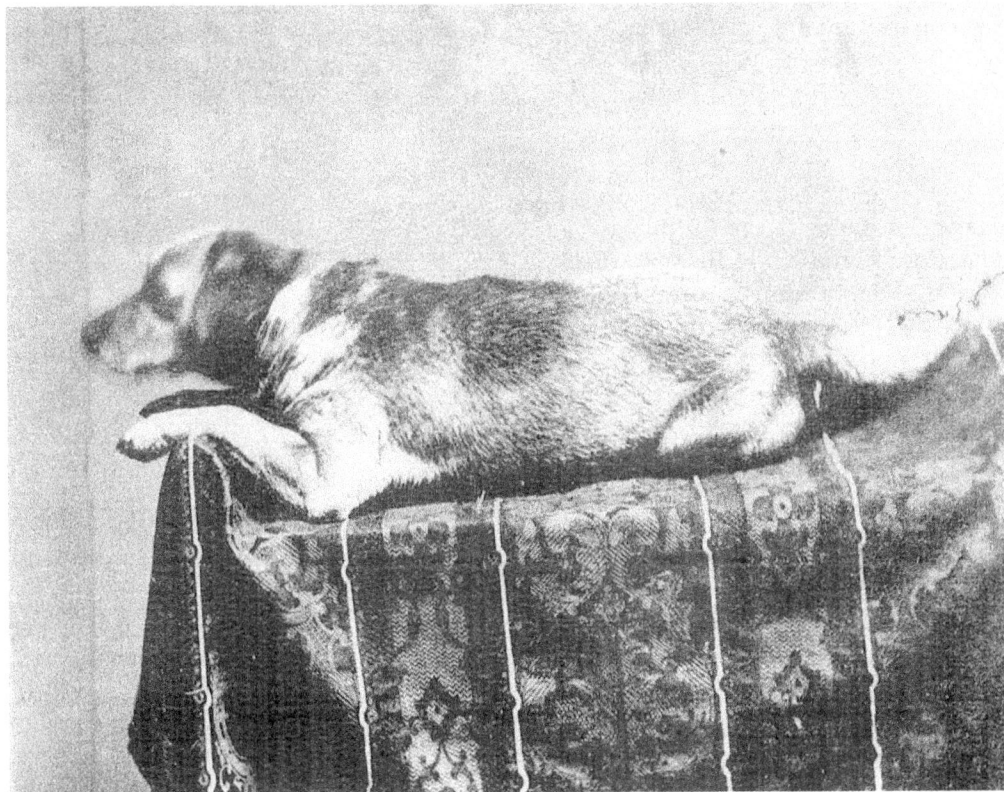

Fido, a yellow mixed-breed dog, was adopted by Abraham Lincoln in 1855. F.W. Ingmire photographed him on his favorite sofa in 1860 (Abraham Lincoln Presidential Library and Museum).

Lorenzo in the Tuscan Province of Florence, Italy, brick worker Carlo Soriani found an injured puppy in a roadside ditch. Carlo and his wife nursed him back to health and named him Fido (meaning "faithful one" from the Latin *fidus*). Fido followed Carlo on his way to the bus stop to see him off to work and would always be at the bus stop to accompany Carlo home. This pattern went on for two years until December 30, 1943, when Carlo was among the thousands of people killed in Borgo San Lorenzo during an Allied bombardment. Every day for fourteen years (more than 5,000 times) Fido continued to meet the evening bus looking for his dear friend, until his own death on June 9, 1958; he was buried next to Carlo. Fido became well-known for his dedication and became a source of media attention. In 1957, the mayor of Borgo San Lorenzo awarded Fido a gold medal and unveiled a statue of the dog in front of Carlo's widow and an onlooking crowd.

Fighting Mac (Dog)

One day in the fall of 1917, Sergeant Major Balcombe (of the 449th Siege Battery, Royal Garrison Artillery, in the Battle at the Somme front near Gouzeaucourt Wood) went to a nearby town for poultry but came back with a retriever-mix puppy. "Mac" was named after the highly respected Sir Hector Archibald MacDonald, a British soldier who had worked his way up the ranks to become a general. Mac was seven weeks old when he was first exposed to war, and from that point on he experienced a lot of action including German shells, bombs and gassing. It came to the point that when he heard a German plane approaching, he would lie flat on the ground, look up to the sky and snarl. After the war, Balcombe was sent to Germany with the Army of Occupation, which unfortunately did not allow dogs. Balcombe had decided to put Mac to sleep,

but when the veterinarian was about to administer the lethal injection, the soldier gave Mac a pass. Mac lived with Balcombe until he died at the age of nine.

Figure (Horse)

Figure was born in Massachusetts in 1789 and then transported to Vermont by Justin Morgan, a music teacher, innkeeper, and town clerk. This small horse of unknown lineage (believed to have been sired by an English Stallion named True Briton) had exceptional intelligence, strength and speed, which was inherited by his offspring and their descendants. At stud, he covered as many as twelve mares per day. After Morgan's death, a series of different owners continued to overwork and neglect Figure, leading to his death around 1821. Twenty years later, many fine little horses, nearly identical in appearance, began attracting attention and were ultimately traced back to Figure. In 1857, the Morgan Horse was declared a new breed.

Filax (Dog)

Filax, a sheepdog not suited for the dog-show circuit, was donated to the Red Cross by his owner. He is said to have saved more than 100 French soldiers traveling over "no-man's land."

Fizo (Dog)

Fizo, an eight-year-old male terrier, was Australia's first recipient of the RSPCA Purple Cross Award, which is the organization's highest award given for animal bravery. Fizo received the award on September 25, 1996, for saving three children, his owner and two friends, from a venomous snake. Fizo jumped off a balcony when he saw the snake close to the children, and held it in his mouth until it died, suffering several bites. Immediately after, he collapsed and was rushed to the animal clinic; he survived to receive his award.

Flash (Dog)

Flash was a Lurcher assigned, along with Boxer the Airedale, to deliver messages for the 34th Infantry Division. Both dogs delivered messages through very deep mud in as little as twenty minutes, whereas it would take a man two hours.

Flush (Dog)

Elizabeth Barrett Browning's pet cocker spaniel, Flush, was the subject of two of her poems, "To Flush, My Dog" and "Flush or Faunus," and is often mentioned in her other writings. Virginia Woolf's *Flush* (1933) is a biography of Flush from his point of view. Flush was given to Elizabeth by Mary Russell Mitford as a gift around 1841 when Elizabeth was ill and bedridden. Flush spent most of the next three years lying at her feet on the sofa in her bedroom. During occasional outings, he was stolen three times, when dognapping was profitable. When Robert Browning began to woo Elizabeth in 1845, Flush bit him. After the couple's marriage, Flush accompanied them to Italy.

Fred (Cat)

Referred to as "The Undercover Cat," Fred, born in May 2005, was a stray in Brooklyn when he was rescued by Animal Care & Control of New York City while battling pneumonia and a collapsed lung. Assistant District Attorney Carol Moran adopted Fred and nursed him back to health. Fred participated in a sting operation involving exposure of a fake veterinarian, Steven Vassall. Posing as a patient, Fred worked with undercover detective Stephanie Green-Jones. Vassall was arrested and charged with unauthorized veterinary practice, criminal mischief, injuring animals, and petty larceny. Vassall pleaded guilty and was given probation and mandatory psychiatric treatment. For his services Fred received a Law Enforcement Appreciation Award on May 18, 2006, by Brooklyn District Attorney Charles J. Haynes. On July 8, 2006, Fred was honored at "Broadway Barks 8!" which is the New York City Theater District's dog and cat adopt-a-thon benefit hosted by Mary Tyler Moore and Bernadette Peters; he received the Mayor's Alliance Award. After the sting operation Fred was training to be a teaching-animal in the District Attorney's Office "Legal Lives" program, which teaches children how to treat and care for animals. On August 9, 2006, at 15 months old, Fred was struck and killed by a car.

Fritz (Horse)

Fritz was owned by film producer Thomas Ince, but he was loved by his co-star, actor William Hart. Hart was the first big star of Western movies and the first to co-star with a horse. During his career, Fritz appeared in eight silent films and was the first horse named in the credits as a co-star. Hart and Fritz performed together from 1915 to 1925 and did all their own stunts. Eventually Hart gained ownership of Fritz. In his autobiography, he describes some of Fritz's risky stunts, such as jumping into moving rivers, through windows, and over fire, and being thrown to the ground after a sudden stop. Fritz received his own fan mail, which often included sugar cubes. After retirement from films, Fritz lived out the rest of his life at Hart's California ranch. Fritz died in 1938 at the age of 31 and was buried at the ranch. A cobblestone monument reading, "Bill Hart's Pinto Pony Fritz—Aged 31 Years—A Loyal Comrade" marks his grave.

Fukutsuru (Bull)

A Wagyu bull in the early 1990s, Fukutsuru was shipped from Japan to

Snake River Farms in Boise, Idaho. He fathered many cattle and aided in the growth of the U.S. Kobe beef market. Even though he died in 2005, his sperm was frozen for long-term breeding.

Fungie (Dolphin)

Also known as the Dingle Dolphin, Fungie is a common bottlenose dolphin residing in the Dingle Harbour and nearby waters off Dingle, Ireland. Locals first spotted him in the harbor in 1983, and he has continued to live in unusually close contact with humans. Dolphins are social animals and often interact closely with each other, but it is rare for them to seek out human contact. Fungie is the first known dolphin to interact positively with humans in the wilds of Ireland. Fungie is known to be playful with swimmers, surfers, kayakers and divers, and there have never been reports of him being aggressive towards humans. A statue of Fungie is located outside the Tourist and Information Office in Dingle. Fungie's story piqued the interest of people asking such questions as, "Do you organize tours or trips to swim with the dolphin?"; "Will he heal my ME/MS/chronic depression, etc.?"; and, "What's so special about Fungie anyway?"

Fup (Cat)

Fup was the resident "Bookstore Cat" at Powell's Technical Books in Portland, Oregon. The store raised money for the Oregon Humane Society by selling gift cards and T-shirts with Fup's image. Fup spent nearly 19 years of her life hiding at the top of the book fixtures at the store. Every morning she would be found sitting in front of the office awaiting her breakfast. Fup became somewhat of a celebrity, receiving gifts, cards, and emails from fans.

Ganda (Rhinoceros)

Five centuries ago German artist Albrecht Dürer, never having seen a rhinoceros, penned the most famous illustration of a rhino in an ink drawing, or woodcut. The subject's story begins in the year 1515 when the governor of Portuguese India, Alphonso Albuquerque, gifted a live rhinoceros to King Manuel I of Portugal. It was quite common in those days for royalty to keep exotic animals. Ganda was captured in what is now the state of Assam and was dispatched with other cargo onto the ship *Nossa Senhora da Ajuda*, which set sail in January from the port of Goa. The ship arrived 120 days later in Lisbon, after sailing westward across the Indian Ocean and Africa's Cape of Good Hope. Ganda's new home was in the royal menagerie at Ribeira Palace, at a distance from the elephants because of the belief that the two species were mortal enemies. To verify this belief, within a few weeks the king arranged a battle between Ganda and his youngest elephant. The elephant was led into the arena and the curtains hiding Ganda were drawn open. An observer, Valentin Ferdinand, wrote that Ganda appeared to be furious and charged violently and broke her chain. The elephant reacted by "uttering a tremendous cry" and running to safety. It is not known how this affected the king; however he re-gifted Ganda to Pope Leo X, and a ship carrying Ganda set sail in December for the Holy City. In January, the ship docked in Marseilles and then headed to Rome. The doomed ship encountered a storm and sunk in the Gulf of Genoa, drowning all aboard, including Ganda. Interestingly Ganda's body was recovered, stuffed and presented to the pope.

Gander (Dog)

Gander, originally named Pal, was a friendly Newfoundland dog raised by the Hayden family in Gander, Newfoundland.

When Gander began to grow too large, the family gave him to the Royal Rifles of Canada Regiment stationed at the Gander International Airport, and he became their mascot. In 1941, the regiment was sent to defend Hong Kong from an enemy invasion. Not only was Gander a mascot but he was also a fellow soldier, scaring away the enemy by barking and nipping at the soldiers' legs. One night in December 1941, Gander sacrificed his life to save the lives of seven soldiers. Gander saw a grenade near a group of wounded Canadian soldiers; he took it and ran away with it and the grenade exploded and killed him. "Sergeant" Gander was never forgotten, and on October 27, 2000, he was awarded the Dickin Medal posthumously, the first to be awarded to a Canadian animal, during a ceremony in Ottawa. The Dickin Medal is thought to be equivalent to the Victoria Cross. The citation reads:

> For saving the lives of Canadian infantry men during the Battle of Lye Mun on Hong Kong Island in December 1941. On three documented occasions, Gander, the Newfoundland mascot of The Royal Rifles of Canada, engaged the enemy as his regiment joined The Winnipeg Grenadiers, members of Battalion Headquarters "C" Force and other Commonwealth troops in their courageous defense of the island. Twice Gander's attacks halted the enemy's advance and protected groups of wounded soldiers. In a final act of bravery, the war dog was killed in action gathering a grenade. Without Gander's intervention, many more lives would have been lost in the assault.

Three Honors:

- Gander's name was listed with those of 1,975 men and two women on the Hong Kong Memorial Wall in Ottawa, Ontario, Canada.
- On July 23, 2015, statues of Gander and his handler were unveiled at Gander Heritage Memorial Park in Gander.
- The Forgotten Heroes Monument at the Cobequid Veterans Memorial Park in Bass River, Nova Scotia, initiated as an elementary school project and designed by 11-year-old Noa Tremblay, commemorates heroic animals. It includes a statue of Gander by Nova Scotian sculptor Clifton Sears.

Gef (Mongoose)

Gef is also referred to as the "Talking Mongoose" or the "Dalby Spook." In September 1931, James, Margaret and 13-year-old daughter Voirrey Irving claimed that behind their farmhouse's wooden wall panels they heard scratching, rustling and voices. The Irvings' farm was located at Cashen's Gap near the hamlet of Dalby on the Isle of Man. Allegedly, from behind the wall panels emerged a creature the size of a rat, with yellow fur and a bushy tail, introducing himself as Gef and telling them it was a mongoose born in New Delhi, India, in 1852. Some of the things Gef reportedly communicated: he was "an extra, extra clever mongoose," an "Earthbound spirit," and "a ghost in the form of a mongoose." The Irvings also claimed that Gef guarded their house, woke them up if they overslept, and chased mice. The story caught the attention of the tabloid press and journalists as well as psychic investigators. In July 1935, Richard S. Lambert, editor of *The Listener*, and paranormal investigator Harry Price investigated the case and wrote a book, *The Haunting of Cashen's Gap* (1936). A hair found at the house was sent for evaluation to naturalist F. Martin Duncan, who concluded it was from a dog, most likely the Irvings' sheepdog. Reginald Pocock of the Natural History Museum evaluated some pawprints and concluded they may have belonged to a dog, but definitely not a mongoose. Some psychic investigators speculated that Gef could be a poltergeist or a ghost, while other investigators concluded that Voirrey Irving used ventriloquism and the parents perpetuated the hoax. James Irving's diaries, as well as reports about the case, are stored in Harry Price's archives in the Senate House Library at the University of London.

Gelert (Dog)

In the city of Beddgelert, Wales, stands a statue of Gelert, the city's most famous dog, known for his extreme loyalty. The "folktale" is written on his tombstone:

> In the 13th century Llewelyn, prince of North Wales, had a palace at Beddgelert. One day he went hunting without Gelert, "The Faithful Hound," who was unaccountably absent. On Llewelyn's return the truant, stained and smeared with blood, joyfully sprang to meet his master. The prince, alarmed, hastened to find his son, and saw the infant's cot empty, the bedclothes and floor covered with blood. The frantic father plunged his sword into the hound's side, thinking it had killed his heir. The dog's dying yell was answered by a child's cry. Llewelyn searched and discovered his boy unharmed, but nearby lay the body of a mighty wolf, which Gelert had slain. The prince, filled with remorse, is said never to have smiled again. He buried Gelert here.

General William Howe's Dog *see* Lila

George (Dog)

George, a nine-year-old, foot-tall Jack Russell terrier, was given an award for bravery by the Society for the Protection of Animals, and Vietnam War veteran Jerrell Hudman sent George one of his three Purple Hearts. George was given these honors for giving his life to save five children in New Zealand from being attacked by two pit bulls in May 2007. George was so severely injured in the fight that he had to be euthanized.

George (Goldfish)

In Australia in 1914, George the goldfish underwent a 45-minute operation to remove a tumor on his head. His owner had two options: to surgically remove the tumor or to euthanize George. His owner was quite attached to him and wanted the surgery. The procedure involved three buckets—one with knockout anesthesia, one with maintenance anesthesia, and one with clean water for recovery. Melbourn veterinarians expected George to live another twenty years. George had been owned for ten years. The operation cost $200.

Geronimo (Beaver)

In the 1940s, Idaho Fish and Game had an abundance of beavers in some areas that were too close to people—their dam-building at times flooded neighborhoods and roads—while in the backcountry there was plenty of space. The usual method of transplanting beavers was trucking or using mules. Neither method was viable because of the hot weather and rough travel. Fish and Game experimented with willow boxes but decided on a box design that would open upon impact when the box hit the ground after being parachuted. But would the beaver die? Geronimo, a plucky male beaver, was to be the test subject. There was a surplus of parachutes from World War II that could be used with the beavers. To test proper heights and box designs, officials dropped Geronimo over and over again. Geronimo finally became resigned, and as soon as they approached him, he would crawl back into the box ready to go. With his help, Fish and Game learned that the best launch height was between 500 and 800 feet to allow the chute to open properly and still be accurate. In 1948, Idaho Fish and Game dropped 76 beavers in the backcountry, with only one fatality. Geronimo's reward for all his work was to be the first male beaver on a first-class airplane seat with three lovely female beavers.

G.I. Joe (Pigeon)

G.I. Joe was one of the most famous American pigeons of World War II, noted for his service in the United States Army Pigeon Service. On October 18, 1943, G.I. Joe saved the lives of the civilians of the village of Calvi Vecchia, Italy, and the British troops of the 56th (London) Infantry Division. A bombing had been scheduled for that day against German positions at the village. G.I. Joe carried an urgent message to alert the 15th Air Force that the 169th (London) Infantry Brigade had captured the village of Calvi Vecchia—just in time to cancel the scheduled bombing. G.I. Joe, flying twenty miles in twenty minutes, saved the lives of thousands of villagers and British troops. On November 4, 1946, Major General Charles Keightley presented G.I. Joe with the Dickin Medal of Gallantry. The citation credits him with "the most outstanding flight made by a United States Army homing pigeon in World War II." He died on June 3, 1961, at the age of eighteen at the Detroit Zoological Gardens and is mounted at the U.S. Army Communications-Electronics Museum at Fort Monmouth, New Jersey. G.I. Joe goes down in history as the first non–British recipient of the medal.

G.I. Joe served in the United States Army Pigeon Service during World War II (Department of Defense).

Ginger (Dog)

Ginger's story is an example of the dogs that suffered from shell shock during WWI. She was able to recover and finish her services as a messenger dog.

Golden Cloud (Horse)

Originally named Golden Cloud, Trigger was a palomino horse born on July 4, 1934, in San Diego, California. Most people remember Trigger as the famous horse in American Western films with owner, rider, and cowboy star Roy Rogers, and also in Rogers' 1950s television series with his wife, Dale Evans, as a buckskin Quarter Horse named Buttermilk. Trigger appeared in 81 of Rogers' films and 101 television episodes. Trigger was so famous he even had a Dell comic book devoted to his many exploits. In the 1938 film *The Adventures of Robin Hood*, Trigger, then named Golden Cloud, began his film career as the mount for Maid Marian, played by Olivia de Havilland. Rogers bought Golden Cloud in 1943 and renamed him Trigger. Besides learning 150 cues, he could also walk 50 feet on his hind legs, sit in a chair, sign his name with an "X," lie down and cover himself with a blanket. Perhaps one of Rogers' best kept secrets was getting Trigger housebroken. Glenn Randall, a wrangler with Hudkins Stables, wrote, "Spending as much time as he does in hotels, theaters and hospitals, this ability comes in mighty handy and it's conceded by most trainers to be Trigger's greatest accomplishment." After Trigger died at age 33 in 1965, Rogers had taxidermist Everett Wilkensen preserve his hide; it was stretched over a foam likeness and mounted in the Roy Rogers-Dale Evans

Trigger, pictured with Roy Rogers, is remembered for starring in American Western films and television programs.

Museum in Apple Valley, California. The museum moved to Victorville, California, in 1976, and a 24-foot fiberglass statue of a rearing Trigger sat on top of the building. In 2003, the mount ended up in Branson, Missouri, the museum's final location. In 2009, the Branson museum closed and Trigger's mounted remains were sold for $266,500 at Christie's auction house in New York City, to RFD-TV, which planned to open a Western museum. Trigger was billed as "The Smartest Horse in the Movies," and in 1949, Trigger's hoofprints and Rogers' handprints were cemented at Grauman's Chinese Theater.

Goldie (Golden Eagle)

In March 1965, sightseers were causing traffic jams by flocking to Regent's Park in London to see a beautiful golden eagle named Goldie, which had escaped from the London Zoo while his cage was being cleaned. Goldie had 12 days of freedom until a zookeeper managed to recapture him by tempting him with a dead rabbit. Goldie returned in good health and reunited with his mate, Regina.

Gordo (Monkey)

Gordo (AKA "Old Reliable"), a South American squirrel monkey, was chosen by NASA to be the first monkey in space. Gordo was chosen for his species' similarities with man's anatomy and their sensitivity to temperature changes. Gordo was launched in the *Jupiter* AM-13 rocket from the Atlantic Missile Range at Cape Canaveral at 3:53 a.m. on December 13, 1958. The flight lasted 15 minutes, with Gordo being weightless for 8.3 minutes and traveling over 1,500 miles to an altitude of 310 miles. Besides a slowing of Gordo's pulse, scientists found no other negative effects on his health. As he returned to Earth at 10,000

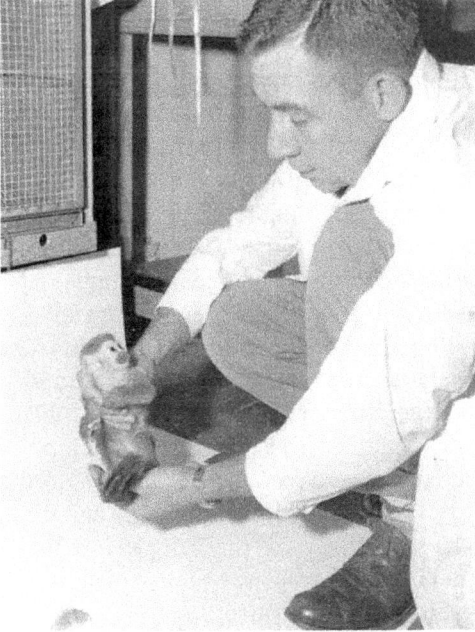

Gordo was chosen by the NASA space program to be the first monkey in space and launched in the *Jupiter* AM-13 rocket on December 13, 1958 (U.S. Army).

miles per hour, sadly his parachute failed to open and both the capsule and Gordo were lost. Gordo may have been alive at impact. Gordo left a legacy for scientists to better understand the effects of space travel on humans. Other monkeys in space have included Able, Baker, Albert, Patricia and Mike.

Grady (Cow)

The farming community of Yukon, Oklahoma, gained fame when Grady, a six-year-old Hereford cow, managed to get herself stuck in a steel-encased grain silo. This predicament took place in the winter of 1949, after Grady had had a difficult birth to a stillborn calf. Being very disoriented, she charged at her owner, Bill Mach, and fortunately he was able to jump out of the way. In her confusion Grady charged through a 17-inch-wide, 25-inch-tall feed hole leading into a silo. So the question was how to free this 1,200-pound cow. The

national news media and dozens of curious onlookers descended upon Yukon. After three days they came to the conclusion that Grady needed to come out the way she got in. Therefore, Ralph Partridge, farming editor for *The Denver Post*, sedated Grady, and then she was doused with 10 pounds of axle grease. A team of men pushed from the back and a team of men pulled from the front—and voila—Grady squeezed through the hole. Well-wishers continued to come to Yukon to pay their respects to Grady. She went on to give birth to several healthy calves before dying from old age in 1961.

Greyfriars Bobby (Dog)

John Gray, a gardener, moved to Edinburgh, Scotland, with his wife Jess and son John. When he could not find work as a gardener he went to work for the local police as a night watchman. Bobby, a Skye terrier, became John's partner and "watchdog." On February 15, 1858, John died of tuberculosis and was buried in Greyfriars Kirkyard. The day after the burial Bobby was discovered sleeping on his master's grave. Despite being evicted he returned night after night. At the firing of the one o'clock gun, Bobby would leave his post to eat his daily meal at the local coffee house. Bobby kept vigil nightly until his own death on January 14, 1872, at age 16. The story generated real interest throughout the world. A life-size commemorative statue unveiled in 1873 opposite Greyfriars Kirkyard, paid for by Baroness Angelia Burdett-Coutts, president of the Ladies Committee of the RSPCA, has become a tourist attraction. The accuracy of the stories of Greyfriars Bobby has been challenged many times. One theory is that Bobby died in 1867 and was replaced with a younger dog. In 1867, the lord provost of Edinburgh presented Bobby with a new collar, which is now on display at the Museum of Edinburgh. The collar has a brass plate inscribed with the words, "Greyfriars Bobby—from the Lord Provost, 1867, licensed." Bobby's headstone reads, "Greyfriars Bobby—died 14th January 1872—aged 16 years—Let his loyalty and devotion be a lesson to us all."

Gua (Chimpanzee)

Not long after Gua was born in Havana, Cuba, on November 15, 1930, she was donated to the Yerkes Laboratories of Primate Biology in Orange Park, Florida, on May 13, 1931. At seven and a half months, Gua went home with scientists Luella and Winthrop Kellogg to be part of a study of cross-rearing to compare the similarities and differences between Gua and their 10-month-old son Donald. For nine months, the two shared the same environmental advantages while the Kelloggs recorded their development. Gua tested ahead of Donald in many aspects, such as reading, learning to walk, and using a cup and spoon. Also physically, Gua was better at jumping, climbing, and using her feet, and as they got older Gua became much stronger. Their recognition of people differed in that Donald identified them by their faces and Gua by their clothes and smell. Also by 16 months of age, Donald began communicating while Gua could not speak. The experiment ended after nine months because Donald started to copy Gua's vocalizations. Gua returned to Florida and died from pneumonia at three years old on December 21, 1933.

Gunner (Dog)

At 10 a.m. on February 19, 1942, the Japanese bombed Darwin, Australia, and a six-month-old Kelpie was found with a broken leg by Leading Aircraftman Percy Westcott. He pulled the dog out of the rubble and took him to the army doctor, who would only treat men with a name and serial number. The doctor ended up setting the leg after Westcott immediately replied that the dog's name was Gunner, serial number 0000. Westcott nursed him back to a full recovery and started to notice some strange behaviors that would prove invaluable. Gunner began whining, pacing, and becoming restless, and as his agitation grew he started barking incessantly up at the sky. Surprisingly, 20 minutes later Japanese warplanes were flying above. Gunner not only continued to repeat these early-warning behaviors, but could also tell the difference between Allied and Japanese planes. These warnings became so reliable that Westcott began to sound a siren at the first sign of the dog's agitation. The warnings gave the troops time to get their planes airborne or take cover, which saved many, many lives.

Gustav (Pigeon)

Also known by his service number NPS.42.31066, Gustav, a grizzle-colored cock pigeon, was part of the Royal Air Force's messenger service during World War II. Gustav was one of six pigeons given to Reuters news correspondent Montague Taylor. On June 6, 1944, Gustav flew 150 miles in five hours from the northern coast of France to Portsmouth to deliver news to the UK with the message, "We are just 20 miles or so off the beaches. First assault troops landed 0750. Signal says no interference from enemy gunfire on beach.... Steaming steadily in formation. Lightnings, typhoons, fortresses crossing since 0545. No enemy aircraft seen." Gustav was presented with the Dickin Medal on November 27, 1944, with the citation reading, "For delivering the first message from the Normandy beaches from a ship off the beachhead while serving with the RAF on June 6, 1944."

Guy (Gorilla)

Guy, a western lowland gorilla, arrived at the London Zoo on November

5, 1947—Guy Fawkes Day—as a tiny baby weighing twenty-three pounds. In 1966 and 1971, Guy weighed 520 pounds, was 5 feet 4 inches tall, had an arm span of 9 feet, upper arm circumference of 23.5 inches, thighs 28 inches thick, and neck 36 inches thick. Guy was one of the London Zoo's most loved animals and was something of a celebrity in the 1960s and 1970s. His gentle nature was what made him so popular. Guy died in 1978 at age 30 from a heart attack during an operation on his infected teeth. In 1961 a statue of Guy was erected in Crystal Palace Park. Arthur Hayward, the Natural History Museum of London's head taxidermist, had the task of modeling and mounting Guy's skin, and Guy was put on display at the museum in November 1982. Also in 1982, a bronze statue of Guy by William Timym was erected near the London Zoo's main entrance. An oil painting by Timym hangs in the library at the Zoological Society of London.

Hachiko (Dog)

Hachiko, an Akita Inu, is known as one of the most loyal dogs in history. Hachiko accompanied his owner, Dr. Eisaburo (also spelled Hidesaburo) Ueno, to the Shibuya station in Tokyo to see him off to work every morning, and he also greeted his return train. In May 1925, Dr. Ueno died during the day at Tokyo University, where he was a professor. That night, Hachiko waited until midnight for his owner's return. For almost nine years, day after day, even escaping from new owners, Hachiko returned to the station to await the return of his deceased owner. After his story appeared in the October 4, 1932, issue of *The Asahi Shimbun*, people began bringing Hachiko food and treats. After his death on March 8, 1935, at the age of 11, his fur was preserved, stuffed, mounted and put on permanent display at the National Science Museum of Japan. In April 1934, a bronze statue of Hachiko was erected at the Shibuya station, and the spot where Hachiko stood for many years is marked with a set of bronze paw prints. The statue was recycled during World War II, but in 1948 a new statue was erected.

Film and book adaptations:

1987 movie *Hachi-ko* (*The Tale of Hachiko*), directed by Seijiro Koyama, tells the story of his life from birth to death, with an imagined spiritual reunion with his master

August 2009 movie *Hachi: A Dog's Tale*, directed by Lasse Hallstrom, starring Richard Gere

1967 children's book *Taka-chan and I: A Dog's Journey to Japan*

2004 children's book *Hachiko: The True Story of a Loyal Dog*, written by Pamela S. Turner and illustrated by Yan Nascimbene

Hachiko, shown here in 1934, was known as one of the most loyal dogs in history for his extreme faithfulness to his owner, Dr. Eisaburo Ueno.

2004 children's book *Hachiko Waits*, written by Leslea Newman and illustrated by Machiyo Kodaira

2008 novel *The Story of Edgar Sawtelle*, written by David Wroblewski

Hachiko's attractions:

- Statue nearby Shibuya station
- Hachiko's stuffed body and old photos at the National Science Museum of Japan
- Hachiko's and Hidesaburo's grave at Aoyama Cemetery
- Hachiko finally meets his master on a new and beautiful monument erected at the University of Tokyo

Haleb (Horse)

Haleb (AKA "Pride of the Desert"), an Arabian horse owned by Homer Davenport, was purchased on August 8, 1900, from Nazim Pasha, the governor of Syria and Aleppo, who in turn had received him from the supreme sheik of the Anezeh. Haleb's mother was the last of the Maneghi Sbeyel mares, whose bloodline is traced back 500 years, and he was sired by a stallion of the family of Sueyman Sebba of the southern desert. Haleb is famous for beating 19 Morgan horses and winning the Justin Morgan Cup in Vermont in June 1907. Haleb died on November 10, 1909, at the age of 8. His skull and partial skeleton were donated to the Smithsonian National Museum of Natural History's Division of Mammals and assigned to the research collection on December 9, 1910.

Ham (Chimpanzee)

Ham (AKA "number 65") was born in 1957 in the then French Cameroons and was eventually captured, ending up in the United States. On January 31, 1961, this space chimp, dressed in a mini space suit, was launched aboard a *Mercury* rocket from Cape Canaveral, Florida. Two years

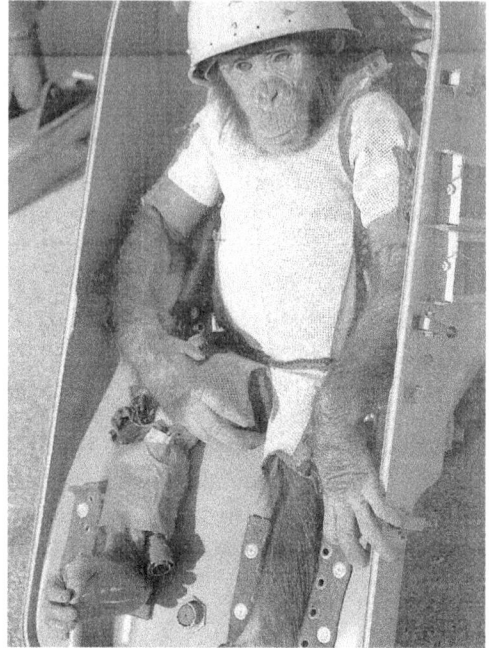

Ham's launch into space paved the way for Alan Shepard's first American human manned flight, May 5, 1961 (NASA).

previously, "number 65" (officials worried that bad publicity might result from the death of a named chimp) had been trained to push a lever within five seconds of seeing a flashing blue light. During Ham's sub-orbital flight everything was going well, except the capsule lost pressure, but Ham's space suit saved his life. Ham performed his tasks during the 16-minute flight and safely returned to Earth. He was renamed Ham, an acronym for the lab that had prepared him—the Holloman Aerospace Medical Center. This mission paved the way for the successful launch of America's first manned flight, with Alan Shepard, on May 5, 1961. Ham became a celebrity and spent the rest of his life at the Washington, D.C., National Zoo and the North Carolina Zoo until his death in 1983.

Hammer (Cat)

In 2004, PFC Hammer, a tiger-striped Iraqi kitten, was made an honorary member of Headquarters Company, 1st

Battalion, 8th Infantry Regiment, 4th Infantry Division in Iraq for his hard work of killing mice, keeping the soldiers' food from being devoured or contaminated. The men were so fond of Hammer that they got help from the animal welfare groups Alley Cat Allies and Military Mascots to bring him to America after their deployment. Hammer lives in Colorado with his comrade, Staff Sgt. Rick Bousfield.

Hanabiko "Koko" (Gorilla)

The name Hanabiko, meaning "fireworks child," refers to Koko's birth date, the Fourth of July (1971). Koko, a female lowland gorilla, was born at the San Francisco Zoo and is known for learning sign language and also for adopting a kitten as a pet. Originally Koko was loaned to animal psychologist Francine Patterson for her doctoral research, but she remained with Patterson, who became her instructor and caregiver. In her reports Patterson claims that Koko had learned more than 1,000 signs in what she terms "Gorilla Sign Language" (GSL), a modified form of American Sign Language (ASL). Also because Koko was exposed to spoken English early on, Patterson reported that the gorilla understood approximately 2,000 words. Other researchers disputed Patterson's findings, arguing that Koko only learned to complete signs because she was rewarded for doing so (operant conditioning). Another concern raised was that the gorilla's signs were left to the interpretation of the handler. Also other critiques suggest that Koko's signing was prompted by her trainer's unconscious cues (The Clever Hans effect). Patterson reported that Koko had demonstrated several complex uses of signs; for example, she used displacement ("the ability to communicate about objects not currently present"). She was also reported to have the abilities to relay memories and use meta-language (defined as using words or symbols for talking about language itself)—for example, "signing 'good sign' to another gorilla who successfully used signing." Patterson also documented Koko inventing new signs to express original thought; for example, Koko did not know the word "ring" yet she signed "finger" and "bracelet"—forming "finger bracelet." Koko was also known for having a pet cat. On her birthday in July 1984 she selected a gray male Manx from a litter of abandoned kittens and named it All Ball. Koko was very nurturing, gentle and loving to the cat. Unfortunately, in December All Ball got out of Koko's cage and was hit by a car and killed. Koko was very sad and could be heard weeping similarly to a human. In 1985, Koko picked out two kittens from a litter, also Manxes, and named one Lipstick for her pink nose and mouth and the other one Smoky because the kitten looked like a cat in one of her books. On her birthday in July 2015, Koko selected another two kittens from a litter and named them Miss Black and Miss Grey. Koko died in her sleep on June 19, 2018, at the age of 46.

The Gorilla Foundation made a statement: "Koko touched the lives of millions as an ambassador for all gorillas and an icon for interspecies communication and empathy," the release said. "She was loved and will be deeply missed."

Quoting the Gorilla Foundation:

> Her impact has been profound and what she has taught us about the emotional capacity of gorillas and their cognitive abilities will continue to shape the world.

Barbara J. King wrote about the BBC documentary *Koko: The Gorilla Who Talks,* when it aired on PBS in 2016:

> Famously, Koko felt quite sad in 1984 when her adopted kitten All Ball was hit by a car and died. How do we know? Here is nonhuman primate grief mediated through language: In historical footage in the film, Patterson is seen asking Koko, 'What happened to Ball?' In reply, Koko utters these

signs in sequence: *cat, cry, have sorry, Koko-love.* And then, after a pause, two more signs: *unattention, visit me.*

In the media Koko and Dr. Patterson's work with her have been the subjects of both books and documentaries:

- 1978 *Koko: A Talking Gorilla,* a documentary film by Barbet Schroeder
- 1980 *Congo,* a novel by Michael Crichton inspired by Koko's story
- 1981 *The Education of Koko,* a book by Patterson and naturalist Eugene Linden
- 1985 *Koko's Kitten,* a picture book by Patterson and photographer Ronald Cohn
- 1986 *Silent Partners: The Legacy of the Ape Language Experiments*, a book by Eugene Linden
- 1987 *Koko's Story,* a children's book by Patterson for Scholastic, Inc.
- 1990 *Koko's Kitten,* a 15-minute re-enactment of the gorilla's adoption of a kitten
- 1999 *A Conversation with Koko,* a PBS documentary for *Nature* (TV series) narrated by Martin Sheen
- 1999 *The Parrot's Lament,* by Eugene Linden
- 2000 *Koko Love!,* a picture book by Patterson and photographer Ronald Cohn
- 2001 *Koko and Robin Williams,* a short featurette on their meeting
- 2016 *Koko: The Gorilla Who Talks to People,* a BBC documentary also shown on PBS

Hanno (White Elephant)

Pope Leo X received this white elephant, named Hanno and probably acquired in India, as a gift on his coronation, from King Manuel I of Portugal. Hanno traveled from Lisbon to Rome in 1514 accompanied by two trainers. Hanno was very well-trained and extremely intelligent, able to understand orders in both Portuguese and Indian. The elephant was also trained to kneel, dance, weep, trumpet, lead parades and entertain at public events as well as parade in important ceremonies in Rome. The pope had a special enclosure built near his residence for his favorite pet. Hanno became ill with constipation just two years after his arrival. The possible cure was a laxative containing gold, which was probably worse than the illness, and he died on June 8, 1516, at age 7. Hanno was buried in the Cortile del Belvedere, and Raphael painted a fresco of him. The pope also wrote a poem dedicated to Hanno and his untimely death (elephants can live up to 70 years). Hanno was commemorated in paintings, poetry, and Latium sculpture. For Romans, he became a symbol of the Orient and for others a symbol of Roman corruption. (In parts of southeastern Asia, white elephants are a rare color for the Asiatic elephant and not a separate species). Also noted is that white elephants "were thought to be sacred, an omen of good fortune and a symbol of all things royal and/or divine" (Listverse website).

Hanno's epitaph written by Pope Leo X:

Under this great hill I lie buried
Mighty elephant which the King Manuel
Having conquered the Orient
Sent as captive to Pope Leo X.
At which the Roman people marveled—
A beast not seen for a long time,
And in my brutish beast they perceived human feelings.
Fate envied me my residence in the blessed Latium
And had not the patience to let me serve my master a full three years.

"The Pope's Elephant," *The Weekly Standard*, by Joshua Gelernter, November 28, 2016.

Harriet (Tortoise)

Harriet was a Galapagos tortoise who had an estimated age of 175 years at

Pen and ink drawing (1514/1516) by Raphael of Hanno, a white elephant belonging to Pope Leo X.

the time of her death in Australia. She was reportedly collected by Charles Darwin during his 1835 visit to the Galapagos Islands as part of his round-the-world expedition, then transported to England, and then brought to her final home, Australia, by a retiring captain of the HMS *Beagle*. There is some doubt about this story because Darwin had never visited the island where Harriet originally came from. Some evidence does support the possibility that she was personally collected by Darwin. Harriet was thought to be male for many years and was initially named Harry. This was changed in the 1960s by a visiting director of Hawaii's Honolulu Zoo. Harriet's favorite food was hibiscus flowers. Harriet died in 2006 of heart failure following a short illness.

Harvey (Dog)

Harvey of the "Barking Dog Regiment" was a white bull terrier with black mask. When his owner, Daniel M. Stearns of Wellsville, Ohio, enlisted in the 104th Ohio Volunteer Infantry in 1862, Harvey became their mascot. There were two other dogs serving with the 104th: Colonel (along with Harvey) was known as a "veteran soldier dog," and Teaser, who joined the regiment in early 1864—thus the reference "Barking Dog Regiment." Lt. Stearns made a brass tag for Harvey's collar, reading, "I am Lt. D.M. Stearns' dog. Whose dog are you?" Some of the letters that the soldiers wrote home were about the antics of the dogs, mainly Harvey. On February 14, 1864, Captain William Jordan wrote to his children about Harvey and Colonel having the run of the regiment, "sleeping wherever they desired or standing guard in the evenings." Captain Jordan also wrote about Teaser chasing a pet squirrel and Harvey picking up the squirrel in his mouth and moving it to safety. In a letter to his brother, Private Adam Weaver wrote that when Harvey was at the campfire sing-alongs, he would often bark and sway from side-to-side. In 1864, while fighting in the Atlanta Campaign, Harvey was wounded and captured but was released the next day. In June 1865, Stearns and Harvey finished their time of service, and presumably Harvey returned home with Stearns; however, his final fate is unknown.

Heather the Leather (Carp)

"Britain's most famous fish," at 50 years old and weighing 52 pounds, was one of the oldest and largest carp in Great Britain. She died in 2010.

Heidi (Opossum)

This cross-eyed North American opossum was found abandoned and then was raised in a wild animal sanctuary in North Carolina. In May 2010, Heidi was given to Germany's Leipzig Zoo by Denmark's Odense Zoo. In 2011, the zoo placed her in a tropical wildlife exhibit along with two other opossums. In her short life Heidi became very popular. After her photograph appeared on the internet, she gained great notoriety, inspiring a YouTube song and a line of stuffed animals. She had over 332,963 followers on Facebook and appeared on *Jimmy Kimmel Live* in a series of pre-taped vignettes instead of appearing on the *83rd Annual Academy Awards*. While appearing on *Jimmy Kimmel Live* she predicted the winners in three Oscar categories. Heidi was euthanized on September 28, 2011, for an unspecified condition.

Helda (Dog)

Helda, an Alsatian sheepdog (German shepherd), was brought to assist at the front lines, and along with Za, another Alsatian sheepdog, she located a German outpost under the guidance of Sergeant Megnin and his assistant. Helda also became a famous sentry dog in the French army when she identified Private Vachet as a traitor. One night when Private Vachet left the camp, Helda and El Tango, another dog, exposed his treasonous behavior; when he denied his activities Helda attacked him into confessing. She was called a canine "Sherlock Holmes."

Hellion (Monkey)

Hellion, a female capuchin monkey, was chosen by Dr. Mary Joan Willard, a psychologist at Tufts-New England Center in Boston, to assist Robert Foster, a quadriplegic, in his home. While Dr. Willard was a research assistant, the idea of using a monkey with grasping ability came to her. She selected capuchins because of their size, alertness, and longevity. Dr. Willard placed her first trainee in 1979; after a year of training by both Willard and Foster, Hellion became a part of Foster's life. Some of the tasks Hellion could perform included feeding Foster, responding to signals he made with a mouth-operated laser, fetching small things, combing his hair, and putting things away.

Hercules (Bear)

Hercules was a grizzly bear purchased as a cub in 1976 from the Highland Wildlife Park in Kingussie, Scotland, by Andy Robin and his wife Maggie. In a year Hercules had grown to 420 pounds. In 1980, Robin made a 60-minute documentary, *Hercules the Wrestling Bear*, to promote their show. Hercules' appearance in Robin's act on the U.K. wrestling circuit led to small acting roles on television. On August 20, 1980, while filming a Kleenex commercial, Hercules escaped from his cage and was missing for 24 days. The fact that he did not eat other animals despite being half-starved endeared him to the public and led to his celebrity status. The "Big Softy" Kleenex advertisements furthered his film career. He appeared in the James Bond movie *Octopussy* in 1983, a Disney documentary, other small film roles and children's documentaries. In 1997, Hercules slipped a disc in his back while filming a BBC documentary, *Eyewitness Bear*, bringing an end to his film career. On February 4, 2001, Hercules died of old age at 25. In 2013, Hercules' adoptive parents unveiled a life-size statue of the bear on the Isle of North Uist. On April 3, 2014, a documentary, *Hercules the Human Bear*, aired on Channel 5 (UK).

Trivia:

- Hercules was featured on the cover of *Time* and helped to promote the Miss World contest
- Hercules caddied for comedian Bob Hope at the Scottish golf course Gleneagles
- Hercules was once named "Personality of the Year" by the Scottish Tourist Board and received a telegram from Ronald Reagan
- The Scottish duo Gaberlunzie dedicated a song to Hercules on their 2003 album *The Travelling Man*
- Hercules was a stunt double for Gentle Ben in a charity wrestling match with Geoff Capes

Herr Olsson and Ushuaia (Cats)

Swedish historian Magnus Forsberg has researched some of the heroic cats that were on the 1901–1904 Swedish Antarctic Expedition on board the ship *Antarctic*, commanded by Captain C.A. Larsen. In October 1902, two cats, Herr Olsson, a male, and Ushuaia, a female, were brought aboard the *Antarctic* at the ship's last port of call in Ushuaia on the southern tip of South America. Ushuaia gave birth to one female kitten, who ended up outliving her parents. When the *Antarctic* sank, the young kitten survived the winter at Paulet Island, living in a hut and on a diet of penguin and fish. The following summer, the Argentine ship *Uruguay* rescued the stranded explorers and the cat. In Buenos Aires, Captain Larsen was awarded a medal by the Argentine Society for the Protection of Animals for rescuing the cat, whose fate is unknown, but she may have been taken to Sweden.

Higgins (Dog)

Higgins was born on December 12, 1957, in Los Angeles, California, but it was 1960 when animal trainer Frank Inn discovered Higgins at the Burbank Animal Shelter. Higgins was thought to be a mix of miniature poodle, cocker spaniel and schnauzer. During a 14-year career in show business, Higgins was best known for being the original Benji and also the uncredited dog who played the role of "Dog" on the sitcom *Petticoat Junction* from 1964 to 1970. Higgins also made guest appearances on the sitcoms *Green Acres* and *The Beverly Hillbillies*. Frank Inn said that Higgins was the smartest dog he had ever trained; he could convey emotions through his facial expressions, could yawn and sneeze, and learned one new trick a week and retained those tricks year after year. In 1971, Higgins starred with Zsa Zsa Gabor and Vincent Price in the film *Mooch Goes to Hollywood*. Around the age of 14, he came out of retirement to star in the 1974 film *Benji*. Higgins' daughter Benjean stepped into the role of Benji beginning with the 1977 film *For the Love of Benji*. Also the dog Mac who played "Tramp" in the series *My Three Sons* was one of Higgins' puppies. Higgins died on November 11, 1975, at age 17. Inn had him cremated and his ashes remained with Inn's daughters after he died in 2002.

Highland Dale (Horse)

Highland Dale was a registered American Saddlebred born on March 4, 1943, in Missouri. He acted in both movies and television from the 1940s to the 1960s. He was discovered at 18 months by horse trainer Ralph McCutcheon, who nicknamed him "Beaut" and trained him to untie knots, play dead and whinny on demand. Highland Dale starred in *Black Beauty* (1946), *Gypsy Colt* (1954), *Giant* (1956), and the television show *Fury* (1955 to 1960). His role in *Giant* made him the biggest horse star of the decade, earning $5,000 a week, and he was insured by the studios for over a quarter million dollars. In 1972, Highland Dale died from respiratory problems.

Homer (Cat)

Gwen Cooper adopted Homer after he was abandoned as a four-week-old stray kitten that had lost both eyes from an infection. Cooper named him after the blind Greek poet who wrote *The Odyssey*. Homer the cat inspired Cooper to write a memoir titled *Homer's Odyssey*. Homer became a fearless feline, taking risks—leaping and climbing around Cooper's apartment. Homer was euthanized on August 21, 2013, at age 16. Cooper donates a lot of the royalties from her books to animal rescue organizations.

Hoover (Seal)

Hoover was a male harbor seal pup found on May 5, 1971, on the shore of Cundy's Harbor, Maine, by Scottie Dunning. Dunning and his brother-in-law, George Swallow, searched for the mother and found her dead on the rocks. Swallow took the seal pup home and when it refused to drink milk from the bottle his neighbor suggested he grind up fish. Hence the pup got his name when he sucked up the ground fish like a Hoover vacuum cleaner. At first Hoover was kept in the Swallows' bathtub, but after a few days he was relocated to a spring-fed pond in back of their house. The Swallows treated Hoover like a family dog, and George and Hoover grew very close. In her book, *Hoover the Seal and George*, Alice Swallow recalls incidences demonstrating their great friendship: "If George was late in giving Hoover his breakfast [he] found his way up the back steps.... George would yell, 'Hello, there.'" Also at times Hoover would hide and would only appear when George yelled, "Get out of there and come over here!" Then Hoover would go and greet George with a big wet kiss. With the seal's increasing growth the Swallows decided to give Hoover to the New England Aquarium. Hoover became famous when it was discovered that he had the ability to say phrases such as, "Hello, there!" and "Come Over Here!" Hoover appeared in *Reader's Digest*, in *The New Yorker*, and on *Good Morning America*. His obituary was published in *The Boston Globe*, and in 2013, a children's book titled *Hoover the Talking Seal*, written by Derek Kaloust and illustrated by Anna Tillett, was published. Hoover died on July 25, 1985, from complications during his annual molt.

Horrie (Dog)

In 1940, Private Jim Moody found a stray Egyptian white terrier in the Libyan desert. Moody adopted Horrie and the dog became an unofficial mascot of the 2/1st Machine Gun Battalion of the Second Australian Imperial Force. Horrie proved to be a valuable member of the Diggers in his ability to detect enemy planes and sound the alarm with his barking, giving the troops time to take cover. Horrie's talent became evident while on a motorbike foray tucked inside Moody's jacket pocket when they were struck by a Stuka aircraft. Moody fell into a ditch and Horrie escaped from the pocket and stood barking at the sky. Upon the battalion's return to Australia in 1942, Moody's commanding officer threatened to throw the dog into the furnace because it was forbidden to repatriate dogs due to the threat of diseases. However, Moody and his company threatened to mutiny; knowing they were serious, the commanding officer allowed Horrie to enter Australia. Moody and Horrie lived together for the next three years; however when they caught the attention of the authorities while raising money for the Red Cross, the authorities ordered that Horrie be destroyed. Did Horrie die? There is some belief that Moody switched a dog from the pound in place of Horrie and that dog was euthanized on March 12, 1945, and that Horrie lived for several more years in Corryong, Victoria. A statue of Horrie was unveiled during the 2016 ANZAC Day service at the Corryong Memorial Hall and Gardens. ANZAC Day is a national day of remembrance commemorating all Australians and New Zealanders who served and died in all wars and is held on April 25.

Books and articles featuring Horrie:

Horrie the Wog Dog: With the A.I.F. in Egypt, Greece, Crete and Palestine, by Ion Idriess, 1948

The Long Carry: A History of the 2/1 Australian Machine Gun Battalion 1939–46, by Philip Hocking, 1997

Horrie the War Dog, by Roland Perry, 2013

Animal Heroes, by Anthony Hill, 2017

"The Dogged Face of Humanity: World War II Dogs," puppy tales by Kerry Martin

"Top 10 Heroic Animal Stories in Australia," by Jackie Nicoletti, *Australian Geographic,* April 13, 2017

Hot Foot Teddy (Bear) (Smokey the Bear)

Hot Foot Teddy was renamed Smokey based on his story. The Capitan Gap fire, a wildfire that burned 17,000 acres in the Lincoln National Forest in the Capitan Mountains of New Mexico, broke out in the spring of 1950. Teddy, an American black bear cub, was found injured up in a tree by firefighters. His paws and hind legs had been burned while he was climbing the tree. There was some dispute as to who actually rescued him. One story is that he was rescued after the fire by a game warden, but the New Mexico State Forestry Division claims that it was a group of soldiers from Fort Bliss, Texas, who had come to help fight the fire. There are also conflicting accounts of who nursed Smokey back to health. *The New York Times* reported in an obituary of Homer C. Pickens, who had been assistant director of the New Mexico Department of Game and Fish, that Pickens had cared for the cub in his home. Other sources, including *Life* magazine, claimed it was New Mexico Department of Game and Fish Ranger Ray Bell who had taken him to Santa Fe, where his family nursed the cub. The national news service picked up his story, making Smokey a celebrity. For the next 26 years, Smokey lived at the National Zoo in Washington, D.C., until his death in 1976. During those years, he received millions of visitors and up to 13,000 letters. Smokey Bear Historical Park was established in New Mexico in 1976 to honor the famous bear. The plaque at his grave reads, "This is the resting place of the first living Smokey Bear ... the living symbol of wildfire prevention and wildlife

Hot Foot Teddy, an American black bear, was the first Smokey bear. He is shown here playing in his pool at the National Zoo (photograph by Francine C. Shroeder, Smithsonian Institution Archives, Record Unit 371, Box 2, Folder December 1976, date c. 1950s).

conservation." Both *The Washington Post* and *The Wall Street Journal* printed obituaries of Smokey.

Hsing-Hsing *see* Ling-Ling

Huaso (Horse)

On February 5, 1949, Huaso set the high-jump record for horses by jumping 2.47 meters (8 feet 1 inch) in Vina del Mar, Chile. His record remains one of the longest-running unbroken sports records in history.

Huberta (Hippopotamus)

Huberta was initially named Hubert until her gender was revealed after her death. In November 1928, Huberta set out on a 1,000-mile trek to the Eastern Cape of South Africa from her home at the St. Lucia Estuary in Zululand. During this three-year journey, Huberta attracted the attention of the general public and journalists, and she became one of the most famous animals in South African history. She was known to cross roads and railroads and walk through towns. She fed herself in parks, gardens, and on farms. For several weeks, she settled at the mouth of the Mhlanga River in Natal, where people visited and gave her treats. Authorities tried to capture her to give her to the Johannesburg Zoo but their attempts failed. Huberta was declared "royal game" by the Natal Provincial Council, thus making it illegal to hunt or catch her. In March 1931 Huberta arrived in East London, South Africa, and one month later she was shot and killed by some hunters, who were arrested and fined. Huberta's body was sent to a taxidermist in London and is now on display at the Amathole Museum in King William's Town. The first report of Huberta, on November 23, 1928, appeared in *The Natal Mercury* and contained the only photograph of her in life. There is a children's book written about Huberta, titled *Hubert the Traveling Hippopotamus*, written by Edmund Lindop and illustrated by Jane Carlson, and published in 1961 by Little, Brown and Company.

Hughes (Donkey)

Hughes was bought by the Canadian Armed Forces engineers in Afghanistan to help transport their equipment over the mountainous terrain of the country.

Humphrey (Cat)

The Chief Mouser to the Cabinet Office was loved by the British people. Humphrey's main job was to rid the Cabinet Office of rats and mice. Being so successful at getting rid of vermin, he served under three prime ministers: Margaret Thatcher, John Major and Tony Blair.

Igloo (Dog)

Igloo's fame in history is for his expeditions to the North and South poles with his master, Admiral Richard Byrd. Igloo (nicknamed "Iggy"), a stray terrier puppy, was found roaming the streets of Washington, D.C., in the winter of 1926 by Maris Boggs. Not able to keep the puppy, she gave him to Byrd to accompany the admiral and his crew on their expedition to the Arctic. Byrd and Igloo made aviation history in 1926 by being the first to fly over the North Pole (the mythical home of Santa Claus), and in 1929, they made the first flight over the South Pole. The two walked down Broadway in a ticker-tape parade upon returning to New York City and also

Igloo pictured with his master, Admiral Richard Byrd, April 12, 1930 (Navy History and Heritage Command).

visited President Herbert Hoover at the White House. Igloo died at the age of six from food poisoning, unfortunately before Byrd could make it back to Boston from his lecturing tour, which he cancelled. *Time* magazine ran an obituary and thousands of letters of condolence were sent to Byrd from children all over the world. Igloo was buried at the pet cemetery in Dedham, Massachusetts, and his grave marker is the shape of an iceberg with a bronze plaque, which reads, "Igloo—He Was More Than a Friend." In front of the courthouse in Winchester, Virginia, there is a life-size bronze statue, sculpted by Dr. Jay Morton, of Byrd dressed in arctic clothing with his right hand stretched out to pat Igloo. In 1931, G.P. Putnam's Sons published a children's book titled *Igloo*, with illustrations by Diana Thorne.

Incitatus (Horse)

Incitatus (the name means "swift" in Latin) was the favorite horse of Caligula, the third emperor of Rome (reigned 37–41 CE). Incitatus was given a marble stable and a furnished house with slaves to wait on the human guests invited to dine with him. He wore a collar made of precious stones. Incitatus was attended by servants and was fed oats mixed with gold flakes. Caligula planned to make Incitatus a consul and appoint him as a priest. The accuracy of his story is generally questioned. It has been suggested that Caligula's treatment of Incitatus was an elaboration by historians and perhaps even satire or an attempt to turn a not-so-great ruler into a villain.

Irma (Dog)

Irma, a German shepherd, began her military career as a messenger dog, but her intelligence and skills led to her switch to being a rescue dog. Irma helped rescue 191 people trapped under blitzed buildings while serving with the London Civil Defence Service during World War II. She also worked with another rescue dog named Psyche, and together they found 233 people, of which 21 were alive. Their owner and handler was Margaret Griffin. Irma had the ability to determine if buried victims were dead or alive—if alive she would bark joyfully and if dead she gave one quick wag of her tail. After one raid she insisted she had found life under a collapsed house and refused to move. Rescuers pulled two young girls alive from the rubble after being trapped for two days. On January 12, 1945, Irma was awarded the

Dickin Medal in recognition of her heroic work. After the war she became a demonstration dog at a dog training school.

J. Fred Muggs (Chimpanzee)

NBC's *The Today Show* with host Dave Garroway premiered on January 14, 1952, and following poor ratings the network decided to add a chimpanzee mascot to the show. That mascot turned out to be J. Fred Muggs, born on March 14, 1952, in French Cameroon, who served as mascot from 1953 to 1957. Muggs' first appearance on *The Today Show* was on February 3, 1953, dressed in diapers, and from that point the ratings improved dramatically. Some of Muggs' antics included sitting in Garroway's lap, mastering more than 500 words, reading the day's newspapers, imitating Popeye, and playing the piano with Steve Allen. The chimp's wardrobe consisted of 450 outfits. Muggs was featured in books, comics, and games. In 1958 one of Muggs' finger paintings graced the cover

J. Fred Muggs, pictured here in 1955, was the mascot for NBC's *The Today Show* from 1953 to 1957.

of *Mad* magazine (number 38). *The Today Show's* producer at that time, Richard Pinkham, estimated that Muggs had raised $100 million for the network. *TV Guide* used to run an annual feature, "The J. Fred Muggs Awards for Distinguished Foolishness," honoring the year's most dubious television programs, episodes, activities and issues.

Jack (Baboon)

Jack was a chacma baboon, pet and assistant to disabled railway signalman James Edwin Wilde, who had lost both his legs in a railroad accident in 1877. Wilde was having trouble getting around on his wooden legs when he came across Jack. In 1881, Wilde spotted Jack leading an oxen cart in his local market. From Jack's owner Wilde learned that the baboon could obey basic commands and push and pull heavy loads. After acquiring Jack, Wilde trained him to pull signal levers and fetch the key to the coal bin according to the number of whistles from an approaching train. At home, Jack pumped water from the well, did other chores, and pushed Wilde to work in a trolley each morning. Some railroad passengers were nervous and concerned about a baboon handling signals. Therefore Jack had to stop until railroad officials tested his abilities—he passed. Jack was officially hired, earning twenty cents a day (about $5 today) and half a bottle of beer on Saturdays. After nine accident-free years on the job, Jack died in 1890 from tuberculosis and was buried next to the signal box.

Jack (Dog)

Handler Errington tells of how his three dogs, Jack the Airedale, and two large Welsh terriers, Whitefoot and Lloyd, worked in Strazeele (in northern France) through difficult roads littered with the remnants of shelling.

In 1881, Jack became the pet and assistant to disabled railway signalman James Edwin Wilde.

Jack (Dog)

Jack's story begins before the Civil War at the firehouse of the Niagara Volunteer Fire Company in Pittsburgh. Jack was a stray brown and white bull terrier puppy adopted by the firehouse after his right leg was broken when he was accidentally run over by a truck returning from a fire. (Another version is that Jack was brutally kicked by one of the firemen.) When the men of the firehouse volunteered in the 102nd Pennsylvania Volunteers Regiment, Jack went with them. Jack also accompanied his soldiers in the Battle of the Wilderness, the Spotsylvania campaigns, and the siege of Petersburg. Jack was captured twice; the first time he escaped on his own and the second time, in May 1863, he was held at Belle Island in Richmond and was exchanged for a Confederate prisoner. On October 19, 1864, at Cedar Creek, Jack was captured for the third time, but was imprisoned for less than a day because the Confederates retreated. In August 1864, while Jack was on furlough, a ball was held in his honor, raising $40 to buy him a silver collar and a medal for his bravery. On December 23, 1864, Jack disappeared in Frederick, Maryland, and was never found. It is speculated that he may have been killed for his silver collar. A portrait of Jack by artist George Prenter hangs in the Soldiers & Sailors Memorial Hall & Museum in Pittsburgh. Florence Biros wrote a book for young adults, *Dog Jack*. There was a movie by the same name starring a deaf female pit bull named Piglet.

Jack (Dog)

Jack's story begins in Front Royal, Virginia, where he lived with his owner, a Confederate jailer.

Because Jack's story was reported in an article in *Harper's Weekly* on November 8, 1862, this young mastiff became one of the best-known mascots of the Civil War. The reporter described Jack to be "of medium size and jetty blackness, except a white breast and a dash of white on each of his four paws." During the war Jack was a Confederate prisoner of war twice: once he escaped in six hours, and the other time he was held for six months until he was exchanged for a Confederate prisoner. The officers told the reporter stories of Jack's brave deeds during the war. While marching to North Carolina, Jack would go ahead to find water for the parched officers, and when he did he rushed back barking to alert them to his find. Another account tells of how Jack would catch chickens and bring them to the soldiers, who were dying from starvation. Yet another story tells of how Jack stayed with an exhausted Union soldier until a wagon came to his rescue. Jack also accompanied his soldiers in the Battle of the Wilderness, the Spotsylvania campaigns, and the siege of Petersburg. Jack was wounded twice but managed to recover on both occasions. In August 1864, while Jack was on furlough, the regiment bought him a silver collar and medal for his bravery.

Jack (Dog)

Jack was a stray German shepherd taken in and cared for by Nurse Edith Cavell in 1911. In the years 1914–1915, Jack provided cover for her in German-occupied Belgium, where she treated wounded Allied soldiers and aided their escape to neutral Holland. Jack was separated from Cavell following her arrest and execution in October 1915. Eventually he found a home with the Dowager Duchess de Croy, whose chateau served as a Red Cross Center on the Franco-Belgian border. Jack lived there another seven years until he died in 1923 and was acquired and preserved by the Imperial War Museum.

Jack Brutus (Dog)

Old Jack was born in Cumberland, Maine, in 1891. Jack became famous in history as the official mascot for Company K, First Connecticut Volunteer Infantry in the Spanish-American war.

Jackie (Baboon)

Jackie was enlisted into the South African Army on August 25, 1915. He became the mascot of the 3rd Regiment of the 1st South African Infantry Brigade. Until then Jackie had been the pet of the Marr family, living on Cheshire Farm in Villieria. Private Albert Marr was allowed to bring Jackie with him into service. Like other soldiers, Jackie drilled and marched

JACKIE   Mascot of the 3RD 5TH AFRICAN INF.

Jackie was enlisted into South African Infantry Brigade on August 25, 1915, and became the mascot for the 3rd Regiment of the 1st South African Infantry Brigade.

with the company and entertained the men. In England he was given the rank of private and a special uniform with buttons, regimental badges and a cap. Jackie proudly wore his uniform, saluting officers, and standing at ease military style; he even used a knife and fork at the mess table. Albert and Jackie were inseparable friends and comrades and fought together for three years on the front lines in France and Flanders. Being able to see, hear and smell better than the soldiers, Jackie often detected the enemy's presence and watched for deadly mines. He marched beside the troops and was devoted to the wounded by remaining by their side. In April 1918, during the Battle of Passchendaele in Belgium, Jackie was wounded in both his arm and leg with pieces of shrapnel. The attending doctor, Lieutenant Colonel R. N. Woodsend, gave Jackie chloroform and was able to dress his wounds and had to amputate his leg. The war ended on November 11, 1918, and Albert and Jackie were shipped to England, where Jackie became a celebrity in the English newspapers. After the war, Private Marr and Jackie helped the Red Cross raise money for wounded soldiers by appearing at fundraisers and taking part in parades. On April 26, 1919, Jackie was officially discharged, receiving a military pension and a civil employment form for discharged soldiers. On July 31, 1920, Jackie received the Pretoria Citizen's Service Medal. Suffering from post-traumatic stress disorder (PTSD or "shell shock"), Jackie died on May 22, 1921, from a heart attack when an extremely loud thunder-clap during a thunderstorm hit close by. Albert Marr passed away at age 84 in Pretoria in August 1973.

Jake Donelson (Rooster)

During war, it seems that mascots have played two very important roles:

giving moral support and performing heroic acts that sometimes saved the lives of soldiers. Although many mascots are dogs or horses, this particular mascot was a rooster named Jake. In spring 1861, Company H, 3rd Regiment, Tennessee Infantry, began training in Cornersville at Camp Cheatham, for service in the Confederate Army. First Lieutenant Jerome B. McCanless saw Jake's unique fighting abilities, and saved Jake from his inevitable fate to be dinner by adopting him on May 25, 1861. The regiment officially enlisted Jake and gave him immunity from danger except for the enemy's bullets. Jake traveled with the regiment until February 16, 1862, when the company surrendered to Grant at Fort Donelson. The soldiers and Jake were captured and after seven months they were exchanged at Vicksburg. The *Civil War Times* reported that while in captivity, McCanless had Jake's portrait painted by "an itinerant painter." Jake died in 1864 and was given a full military burial. McCanless went on to fight in the Vicksburg, Atlanta, and Nashville campaigns, and he died in 1906. Jake's portrait was returned to the McCanless family in Cornersville.

Jean (Dog)

Jean, a border collie known as the "Vitagraph dog," starred in 16 to 18 silent films for Vitagraph Studios. Jean was owned by Laurence Trimble (1887–1954), who, being an aspiring actor and writer, moved to New York in 1906. Their careers began when Trimble wrote an article about the art of movie-making. By chance, the two were on the set of a Vitagraph Studios movie, and it happened that the director was looking for a dog to play opposite Florence Turner (known as the "Vitagraph Girl"). Trimble suggested Jean for the role, thus starting Jean on her way to being a star. Trimble directed many of Jean's films, mostly involving dog adventures, with

Jean starred in 16 to 18 silent films for Vitagraph Studios (*Motion Picture Story Magazine* 1, no. 6, July 1, 1911, p. 1).

Jean playing the role of canine companion. Some of their movies include: *Jean Goes Foraging* (1910), *When the Light Waned* (1911), *Tested by the Flag* (1911), *Jean Intervenes* (1912) and *Playmates* (1912).

Jenny *see* **Titanic** Animals

Jet (Dog)

Jet (full pedigree name Jet of Iada) was a black German shepherd born in July 1942 at Mossley Hill in the Iada kennel of Mrs. Babcock Cleaver. At nine months old he began his training in anti-sabotage work, and at eighteen months he trained in search and rescue. He and his handler, Corporal Wardle, were the first handler and dog to serve in Civil Defence rescue duties. Historian Ian Kikuchi recalled several of Jet's demonstrated acts of bravery:

"Even when searching piles of the remains of factories full of dangerous chemicals and poisonous smoke, Jet's incredible sense of smell was still able to detect survivors."

"While most animals have a natural fear of fire, Jet sometimes had to be held back by his handler from entering burning buildings where the flames were too fierce to enter."

"On one occasion he was searching the ruins of a hotel in Chelsea which had already been thoroughly searched. The ruins had been investigated and dug up several times, but Jet seemed to insist that there was more searching to do. He was particularly interested in a huge section of a wall which was leaning over but hadn't quite fallen down. The ruins were very unstable and eventually one of the men very carefully climbed up to the top, where he found an elderly lady who had been trapped on a brick ledge."

After the war, Jet returned to Liverpool and was presented with the RSPCA's Medallion of Valor for rescuing people from a pit explosion in Cumbria. Jet died in 1949 and there is a memorial honoring him in the flower garden at Calderstones Park.

Jiggs (Dog)

Jiggs, an English bulldog, made history shortly after World War I by becoming the first dog mascot of the United States Marine Corps. Afterwards he had quite a career in the Marines. On October 14, 1922, Jiggs was enlisted into the Corps by Brigadier General Smedley Butler. Private Jiggs was stationed at Quantico, Virginia, where he was promoted to corporal. On January 1, 1924, Jiggs continued his advancement by being given the rank of sergeant, and seven months later was made sergeant major. Jiggs starred in the 1926 movie *Tell It to the Marines* with Lon Chaney. Jiggs died on January 9, 1927; he was mourned throughout the Corps and was buried with full military honors. To this day, the USMC continues the tradition of owning an English bulldog.

Jim *see* Beautiful Jim Key

Jim (Horse)

The death of a young girl named Bessie Baker and her two siblings in St. Louis in 1901 marked the beginning of one of the first safety disasters in the history of American public health. Within a week of being injected with a diphtheria antitoxin, the three children died of tetanus. In the 19th century, it was common to treat diphtheria patients with an antitoxin made from horses' blood. Jim had been a former milk wagon horse before he was used to produce serum containing diphtheria antitoxin in 1901. During his life, Jim produced over 30 quarts of the serum. Unfortunately, Jim contracted tetanus and was euthanized. Bessie's and her two siblings' deaths, as well as the deaths of twelve more children, were traced back to Jim's contaminated blood. These tragedies as well as another contaminated smallpox vaccine spurred the passage of the Biologics Control Act of 1902, thus establishing the Center for Biologics Evaluation and Research, further setting a precedent for regulation of biologics and finally leading to the formation of the U.S. Food and Drug Administration (FDA) in 1906. Jim's unfortunate death got the ball rolling for future regulations in the preparation of many effective vaccines.

Jim the Wonder Dog (Dog)

On March 10, 1925, Jim was born in a litter of seven Llewellin setter puppies in Louisiana. Jim was the runt of the litter and was given as a joke to a fellow hunter, Sam Van Arsdale, in West Plains, Missouri. The joke was on that fellow hunter because Jim "The Wonder Dog" became famous in history for his incredible hunting skills, intelligence and psychic abilities. At first, Jim's hunting skills were questionable when he lay in the shade watching the trainer with the other dogs. However it appeared that he was learning because when he went quail hunting for the first time he knew exactly what to do. During his travels across the country, Jim fetched more than 5,000 quail, more than any other dog. The magazines *Missouri Life* and *Missouri Conservationist* deemed him "The Hunting Dog of the Century." Jim also proved to be very intelligent in numerous ways. He was able to pick out objects and colors, such as a red car or a woman in a blue dress. He was also able to answer questions; for example, in front of a crowd of people Van Arsdale asked Jim, "What would I do if I had a stomach ache?" and amazingly Jim nudged the town physician. Jim could also understand commands in other languages. For example, a man in the same crowd asked Jim in French if there was a Bible in the crowd, and the dog nudged a minister who had a Bible in his coat pocket. He could also understand other languages including Spanish, Italian, German and Greek. Jim could also understand shorthand and Morse code. As Jim's story spread, he was known as "Jim the Wonder Dog" and was featured in *Ripley's Believe It or Not*. Another one of Jim's talents was predicting future events. Some examples include predicting the gender of unborn babies and the winners of the 1936 presidential election, World Series and Kentucky Derby. After many examinations by veterinarians and psychiatrists, the doctors concluded that Jim possessed an occult power. On March 18, 1937, at the age of 12, Jim passed away. Jim is buried in the Ridge Park Cemetery in Marshall, Missouri. On his gravestone is written, "Since Jim was smarter than most people in here, anyhow." Caretakers say that Jim's is the most visited grave in the cemetery. In 1999, a small-town garden was dedicated to Jim.

Jimmy (Donkey)

The tale of Jimmy (AKA Neddy) begins when he was born in France during the Battle of the Somme in 1916. It is said that he was adopted and became the mascot of the Cameronian Scottish Rifles. He served the next two years in the trenches, carrying wounded soldiers to safety and transporting supplies. Jimmy was an honorary sergeant with three stripes on his bridle. After the war he was sold to a Mrs. Heath, and over the next 23 years he appeared at many charity events and raised thousands of pounds for the RSPCA. He died in 1943 and is buried in Peterborough's Central Park. There is some speculation that the tale is just a hoax. In the 1970s, George Walding's son, Derek, a horse trader, revealed that his father had bought Jimmy at a horse sale in Southampton. Being that the prices for horses were too high, he bought Jimmy from a group of gypsies. He tried to sell him to Sanger's Circus but they did not want him. Then George Walding tried to sell him to the Peterborough Cattle Market, using the World War I story as an incentive. Eventually Jimmy was bought by an RSPCA inspector. War historian and Cameronian Scottish Regimental Association member Sam Morrell disputes this story and says that Jimmy is a large part of Cameronian history. Jimmy's story was also featured in an old three-page comic strip called *Jimmy of the Cameronians— The Four-Footed Hero of the Somme*. It is said that he was trained to salute the soldiers with one hoof.

A memorial stone was laid with the following inscription:

Our Jimmy
Born on The Somme June 1916
Mascot of the 1st Scottish Rifles

Died 10 May 1943
Bought by Mrs. Heath in 1920
To give him a good home
And to promote interest in the RSPCA

Jiro *see* **Taro and Jiro**

Jock (Cat)

Jock, a marmalade cat, was a gift to Winston Churchill for his 88th birthday from his private secretary and friend Jock Colville. Jock seemed to be one of Churchill's favorite pets, accompanying him everywhere. Jock went with Churchill to his London home at Hyde Park Gate, traveled with him back to Chartwell, and in July 1964, Churchill and Jock were photographed leaving Churchill's London home for Parliament. Jock was so admired that the Churchill family deemed that there should always be a marmalade cat in residence at Chartwell. Jock VI, a marmalade cat, is keeping Jock's legacy alive.

Joe (Crow)

"Joe the Crow" was one of many animals serving alongside Canadians during the Korean War (1950 to 1953). Joe was the mascot of the naval training base HMCS Cornwallis in Nova Scotia. Joe's function was to keep the sailors company.

Jofi (Dog)

For almost all of his life, Sigmund Freud was not fond of dogs unless they appeared in his patients' dreams. In 1903, when Freud was in his mid–70s, he received two red chows. One of the dogs, a fluff ball named Jofi, became Freud's special companion. Jofi became his right-hand canine and accompanied him during patient sessions. Thus Freud began to think that dogs had a calming effect over people. He also believed that dogs

have the ability to read people's emotional states and are great judges of character. Jofi would lie relatively near a patient if he was calm, whereas she would keep her distance if the patient was anxious. Jofi also had an internal clock that would alert her when Freud's sessions were over. After exactly 50 minutes, she would get up, stretch, and head for the door. Freud described Jofi as a charming creature. Jofi died in 1937 and despite being devastated over her loss, Freud acquired another chow named Lun.

Jonathan (Tortoise)

After the death of Harriet, a 175-year-old giant Galapagos land tortoise, in 2005, Jonathan, now more than 190 years old, is now thought to be the oldest land animal. In 1882 at age 50, Jonathan, along with three other tortoises, was taken from the Seychelles to the island of St. Helena, where Jonathan now lives at the home of the governor. As Jonathan was aging, he began to slow down, with declining eyesight and sense of smell. He struggled to eat, grabbing at twigs, leaves and dirt for food. Dr. Joe Hollins, the island's vet, restored Jonathan to health by feeding him a high-calorie and nutritious diet of apples, carrots, cucumbers, bananas and guava. Dr. Hollins said, "His life has been transformed. The life expectancy of a giant tortoise is 150 but there is no reason why Jonathan won't still be here after we have all gone." Despite Jonathan's good health, he is too old for breeding to help the dwindling numbers of the Seychelles giant tortoise.

Joni (Chimpanzee)

Joni was raised by Nadezhda Ladygina-Kohts in her Moscow home from 1913 to 1916. She had wanted to observe, test, and record the chimp from the age of 18 months to four years. After

Jonathan, now more than 190 years old, is thought to be the oldest land animal.

approximately three years under study, Joni died of a respiratory illness in 1916. In 1925, Kohts' son Roody was born. Around 1929, Kohts began work on a comparative study based on detailed descriptions in her diary of Joni and Roody. In 1935, Kohts published *Infant Chimpanzee and Human Child*, which focuses on her comparative study between the behavior of an infant chimpanzee and a human child. The ingenious experiments conducted by Kohts were considered ahead of their time.

Josiah (Badger)

President Theodore Roosevelt met Josiah in Kansas during his 1903 railroad tour of the American West. Rumor has it that a twelve-year-old girl named Pearl Gorsuch approached him and asked if he would like a badger. Roosevelt took the baby badger from Pearl's father, Josiah. He was delighted when the badger began to nibble on his finger. Now named Josiah, or "Josh" for short, the badger left Kansas on the train, resting in his own cage and being fed milk and potatoes. He was very good-natured and lived in a sturdy cage to protect him from the other pets. When Josiah was let loose, he got into trouble, often for nipping at legs but never faces. Josiah became less good-natured and found a new home at the Bronx Zoo in New York City. The Roosevelts came to visit him.

Journey *see* OR-7

Judy (Dog)

A great deal on Judy's story is from memories, diaries, and ship's logs collected by Vic Oliver, Les Searle, George White and William Wilson and detailed in *The Judy Story*. Judy, an English pointer with liver and white coloration, was born in 1936 in a kennel in Shanghai, China, and was purchased by Lieutenant Commander J. Waldergrave and Chief Petty Officer Charles Jefferey. Judy, or "Judy of Sussex," spent most of her war service as the mascot of the Royal Navy gunboats the HMS *Gnat* and the HMS *Grasshopper*, patrolling the Yangtze River. Judy had an important job: to alert the sailors of river pirates, hostile Japanese aircraft, and dangerous animals. However, she is perhaps best known for being the only dog to be registered an official prisoner of war of the Japanese during World War II (number 81A Gloergoer, Medan). She was awarded the Dickin Medal for outstanding animal bravery. In 1942, during the Battle of Singapore the *Grasshopper* was bombed en route to the Dutch East Indies, leaving the crew on a deserted island in the South China Sea with no food or water. Judy saved everyone's life by sniffing and unearthing a fresh water spring. Managing to get to Sumatra, the crew traveled 200 miles across the island through jungle for five weeks, coming to Sawahlunto, where they caught a train to Padang. They missed the evacuation ship by nine days and became prisoners of the Japanese. It was at the Medan camp in Indonesia where Judy met leading aircraftsman Frank Williams, and the two became inseparable friends. While at the camp Judy had a litter of puppies. Of the five surviving puppies, one was given to the camp commandant; one was smuggled into the women's camp; one was given to the Red Cross in Medan; a drunk guard beat one to death; and the last one remained in the camp. On June 26, 1944, the POW ship SS *Van Warwyck* was torpedoed by the HMS *Truculent*. Frank and Judy both escaped but were separated. Frank was recaptured and sent to another camp not knowing if Judy survived. Judy was found in the water by other survivors, and at the dock, Les Searle attempted to smuggle her aboard a truck with him. However, the dog was discovered and a Japanese captain threatened to kill her until the former commander of the Medan camp overrode the order and allowed her to travel to a new camp, where Judy and Frank were reunited. After four weeks they were transferred to Sumatra, where Frank spent a year laying railroad track and eating maggot-ridden tapioca. On February 17, 1950, at 5:00 p.m. (Tanzanian time), Judy succumbed to a tetanus infection and was euthanized. Judy's body was wrapped in her Royal Air Force jacket and laid in a wooden coffin. She was buried near the home she made with Frank in Nachingwea, and her grave is topped with a plaque reading:

> In Memory of Judy D.M. Canine V.C.
> Breed English Pointer
> Born Shanghai February 1936, died February 1950.
> Wounded February 14, 1942.
> Bombed and sunk HMS *Grasshopper*.
> Lingga Archipelago February 14, 1942.
> Torpedoed SS *Van Waerwijck*.
> Malacca Straits June 26, 1943.
> Japanese Prisoner of War March 1942– August 1945.
> China Ceylon Java England Egypt Burma.
> Singapore Malaya Sumatra E Africa.
> They Also Served.

Judy's citation for the Dickin Medal reads:

> For magnificent courage and endurance in Japanese prison camps, thus helping to maintain morale among her fellow prisoners and for saving many lives by her intelligence and watchfulness.

Some other sources of interest on the Japanese war in China include:

The journals of Gould Hunter Thomas about his three years in Shanghai collected in his book *An American in China: 1936–39*.

Some contemporary sources on the Japanese takeover include:

Thomas' book mentioned above
Edgar Snow's *Red Star Over China*
Quotes gathered by English writer John Gittings at johngittings.com
Iris Shang's book *The Rape of Nanking: The Forgotten Holocaust of World War II*
The Naval History and Heritage Command website (history.navy.mil)
Information about war dogs in World War II can be found in Anna M. Waller's 1958 book *Dogs and National Defense*

Frank Williams' background mainly comes from his memorial website his family prepared: frankwilliams.ca.

Frank Williams' escape to the docks is recorded in:

Stanley Saddington's *Escape Impossible*
Arthur Donahue's *Last Flight from Singapore*
Geoffrey Brooke's *Singapore's Dunkirk*

Accounts of Judy's water-finding heroics appear in the September 1975 issue of *Look and Learn*, a young adult British periodical.

The website Uboat.net details the submarine HMS *Truculent* torpedoing of Frank and Judy's POW transport.

Jumbo *see* Peggy

Jumbo (Elephant)

Jumbo, "the world's most famous elephant," was born in East Africa in 1860. When he was a year old, he was captured in Ethiopia and survived a trek to Europe alongside other captured animals including ostriches, camels, antelopes and giraffes. Upon arriving in Italy in 1862 he went to a zoo in Paris, France, and then was sold to the London Zoo. He was named Jumbo, most likely from one of two Swahili words: *jumbe*, meaning "chief," or *jambe*, meaning "hello." London Zoo animal caretaker Matthew Scott met Jumbo when he was five years old. He had been badly mistreated at the Paris zoo, and Scott vowed to nurse him back to health. With his special care, Jumbo grew from four hundred pounds and five-feet tall to fourteen thousand pounds and eleven-and-a-half-feet tall by 1880. Abraham Bartlett, superintendent of the London Zoo, sold Jumbo for $10,000 to P.T. Barnum, James A. Bailey and James L. Hutchinson of the Greatest Show on Earth fame. In March 1882, Jumbo began his long journey to New York and across North America accompanied by Matthew Scott, delighting millions of circus fans. Jumbo was hit by a train engine and died on September 15, 1885, in Saint Thomas, Ontario, Canada. Barnum had Jumbo's skin stuffed and it wound up at Tufts University in Boston, Massachusetts. A life-size statue of Jumbo was erected in 1985, where he had died. Jumbo's legacy lives on in various forms of media:

A children's book written by Helen Aberson in 1939, titled *Dumbo*, tells about a circus elephant named Jumbo Jr., who was treated cruelly until it was discovered that he could fly with his giant ears.

In 1941, Walt Disney produced a full-length animated film of Helen Aberson's book *Dumbo*.

In 1962, there was a Doris Day musical, *Jumbo* (based on a Broadway show), about a large elephant at the center of a battle between circus owners.

A book written by Peter and Connie Roop and illustrated by Zachary Pullen, *Tales of Famous Animals*, was published by Scholastic, Inc., in 2012 (pages 18–21).

Jungle Joe (Pigeon)

Jungle Joe was only 4 months old when he began flying for the military

Jumbo was famous for being the main attraction in P.T. Barnum's Greatest Show on Earth (poster ca. 1882).

during World War II. Having lost their radio operator in the jungles of Burma, the military's only source of communication with headquarters was Jungle Joe. For a week, the patrol collected vital information, and on the seventh day they released Joe with the information to headquarters. Joe flew 225 miles over high mountains to deliver his message, resulting in the capture of a large part of Burma. The message also revealed the location of the patrol, and the military was able to parachute another radio operator to their location.

Just Nuisance (Dog)

Just Nuisance, a Great Dane, is generally believed to be born on April 1, 1937, in Rondebosch, Cape Town. He was sold to Benjamin Chaney and accompanied him to Simon's Town, where Chaney was to run the United Services Institute (USI). This was where Nuisance began to follow the sailors of the British Royal Navy back to their ships based in the harbor. The Great Dane was lovingly nicknamed Nuisance because he would block the gangway and the sailors would often say to him, "You're just a nuisance." Just Nuisance served between 1939 and 1944 at HMS Afrikander, a Royal Naval Base in South Africa. He was enlisted on August 25, 1939—the only dog to be officially enlisted in the Royal Navy. His surname was "Nuisance"; his trade was "Bonecrusher"; and his religious affiliation was "Canine Divinity League." He was promoted from the rank of ordinary seaman to able seaman, which entitled him to free rations. He never went to sea but attended promotional events and often accompanied drunken sailors back to base after the pubs closed. Just Nuisance became very popular, boosting morale, and roamed freely—sometimes even taking day trips by train. His union with a Great Dane named Adina produced five puppies. Two, named Victor and Wilhelmina, were auctioned off at Cape Town, raising funds for the war effort. Postcards of Nuisance and Adina were sold to support the Naval Wartime Welfare Fund. Sadly Nuisance was involved in a car accident in 1944, which led to his death at the age of seven. He was buried at Klaver

Camp with full naval honors, including a gun salute and the playing of "The Last Post." There was a statue of him erected in Simon's Town, and the Simon's Town Museum has an exhibition dedicated to his life. Since 2000, there has been a Just Nuisance Commemoration Day Parade of Great Danes from which one dog wins the Just Nuisance look-alike contest. Terence Sisson, an English navy man, wrote a booklet, *Just Nuisance AB: His Full Story*; the 8th edition was published in June 2001. Leslie M. Steyn also wrote a booklet, *Just Nuisance: Life Story of an Able Seaman Who Leads a Dog's Life*. And *Able Seaman Just Nuisance: Based on a True Story*, written by Sherri Rowe, was published in May 2015.

Kabang (Dog)

This shepherd-mix Askal was adopted by Rudi Bunggai as a stray puppy in the Philippines in 2011. Bunggai's 9-year-old daughter and 3-year-old cousin were attempting to cross a busy street in the path of a motorcycle. Seeing the girls in danger, Kabang jumped at the motorcycle, knocking it over. If not for Kabang, the girls would have been seriously injured or killed. Kabang got caught in the front wheel and her upper snout was crushed. She ran away but returned two weeks later. She suffered severe injuries, but Rudi refused to have her euthanized. The dog adapted because the family could not afford surgery. In 2012, the story became known on the internet and an aid campaign was launched by a nurse, Karen Kenngott, called "Care for Kabang." Kabang was brought to UC Davis Veterinary Medical Teaching Hospital in the United States, where she received treatment and surgery. Veterinarians also discovered that she had cancer and treated her with chemotherapy. The $27,000 cost of the surgery and treatment was paid for by donations from 47 countries. Kabang returned home a hero. Veterinarian Anton Lim remarked, "[Kabang is] as normal as she can be. She doesn't need any special medication. So aside from the aesthetic, she's normal."

Kaiser and Max (Dogs)

Kaiser and Max were English poet Matthew Arnold's dogs in 1882, when he wrote "Poor Matthias," an elegy to his canary. In the poem, Arnold also recalled other pets—Arossa the cat, Geist the dachshund, and another dog—then wrote a poem about his present dogs.

For his canary:

Poor Matthias

Poor Matthias—Found him lying
Fallen beneath his perch and dying?...

For Kaiser and Max

Max a dachshund without blot—
Kaiser should be, but is not
Kaiser with his collie face
Penitent for want of race

Max, with shining yellow coat
Prinking ears and dewlap throat
Kaiser with his collie face,
Penitent for want of race

Arnold did not think he would write about either dog, but in 1887 he wrote his last poem.

Kaiser Dead

What, Kaiser dead? The heavy news
Post-haste to Cobham calls the Muse...

Yes Max and we grow slow and sad;
But Kai, a tireless shepherd-lad,
Teeming with plans, alert, and glad
In work or play,
Like sunshine went and came, and bade
Live out the day!...

Kaja Boy (Pigeon)

Kaja Boy was also known as "the little streak" for successfully delivering his message despite being exhausted and wounded. According to reports, he made a full recovery from his injuries.

Kandula (Elephant)

Kandula, captured in the forest of Ceylon (today's Sri Lanka), was the royal mount of King Dutthagamani, ruler of Ceylon during the second century BCE. During the siege of Vijitanagara (161 BCE), Kandula performed heroically, despite being wounded by molten black pitch. It was after King Elara (South Indian rival) was defeated that Kandula became King Dutthagamani's mount.

Karal (Lion)

Montpelier aristocrat Baron Richard d'Arcy was attacked and killed by his two-year-old pet lion, Karal. It seems he was trying to get the lion to go into the bathroom where the animal usually slept.

Karvardi (Bull)

Karvardi was the champion of Nelore.

Keiko (Orca)

Known earlier as Sigi and Kago, Keiko, a male orca, was born in 1976. He is most well-known as the orca portraying Willy in the 1993 film *Free Willy*. In Japanese, Keiko means "lucky one" or "blessed child." Keiko's story begins when he was captured near the coast of Iceland in 1979 and sold to an Icelandic aquarium. Keiko started to perform for the public in 1982 at Marineland in Niagara Falls, Ontario, Canada, where he developed skin lesions and was bullied by the older orcas. In 1985, he was sold to Reino Aventura (now Six Flags Mexico) in Mexico City (and there was named Keido). Being only ten feet long, he was placed in a tank made for dolphins. It was at this amusement park where he was found by movie scouts for *Free Willy*. In February 1995, from donations from children around the world, Warner Brothers, The Humane Society and billionaire Craig McCaw, the *Free Willy-Keiko Foundation* was established. After the organization raised more than $7 million from donations, Keiko was given to the Oregon Coast Aquarium in Newport in 1995. A rehab tank was built for him, and upon arriving in Oregon in 1996, he came home to a new 2-million-gallon concrete enclosure filled with seawater, where he regained his strength. A plan for returning Keiko to the wild spurred a great deal of controversy. Some felt that returning him to the wild would most certainly fail due to his many years in captivity. Some suggested that he be monitored with radio and satellite tags to make it possible to ensure his safety and to return him to human care if needed. Norwegian pro-whaling politician Steinar Bastesen said the orca should be killed and the meat sent to Africa for foreign aid. Keiko was released in 2002 but, despite training to prepare, he died of pneumonia on December 12, 2003, in Taknes Bay, Norway, at 27 years of age. Keiko's failure to survive was attributed to his capture at an early age, his long history in captivity, and his strong bonds with humans. Others felt the release had been a success. *The Huffington Post* called it a "phenomenal success … giving him years of health and freedom." About $20 million was spent freeing Keiko. Lori Marino of the Whale Sanctuary Project suggested that an alternative to freeing captive whales may be a "sanctuary" or "oceanic enclosure." In a 2016 interview, Marino said, "It's a solemn responsibility, and it's the best we can do for animals that are in captivity."

Ketzel Cotel (Cat)

Ketzel, a tuxedo cat, leapt to fame one morning when Professor Cotel was doing his piano exercise in Bach fugues when she jumped and worked her way down the keyboard. The professor transcribed her music, which he described as "structural elegance." The professor nominated Ketzel's piece of music to the 1997 Paris New Music Review for works under 60 seconds. Her work, titled "Piece for Piano, Four Paws," was awarded "special mention" and was said to be "quite well written." This was her only piece of music. Ketzel is mentioned in the book *The World of Women in Classical Music*.

Khan *see* **Rifleman Khan**

Kiddo (Cat)

Walter Wellman (1858–1934), an American explorer, aeronaut, and journalist, looked to three newspapers—*The New York Times*, *The Chicago Record-Herald* and Britain's *Daily Telegraph*—for financing his flight over the Atlantic Ocean in his airship, *America*. Kiddo, the "airship cat," was one of a pair of twin stray cats that lived around the hangar. Kiddo's brother had died a few weeks before the flight, so Kiddo became the first cat to take flight over the Atlantic Ocean in an airship. Kiddo belonged to one of Wellman's crew members and the crew referred to him as "Kitty." On October 15, 1910, five men and one cat took off from Atlantic City, New Jersey. A storm from the northeast blew them off course; along with engine difficulties, it meant that the crew and cat had to take to a lifeboat. The crew and Kiddo were rescued by the British steamship RMS *Trent*, which was en route from Bermuda to New York. Despite not crossing the Atlantic, the *America* had remained in the air for 72 hours and had flown over 1,000 miles. After being rescued Kiddo was renamed Trent after the rescue ship and gained somewhat of a celebrity status. The cat was displayed in the window of Gimbel's, one of the leading department stores, and also had its photo reproduced on a post-card. Kiddo retired from aviation and went to live with Wellman's daughter Edith.

Kidron (Horse)

Warhorse Kidron was ridden by General John J. Pershing in some of his early military campaigns but became famous during World War I when he was used mainly as a ceremonial animal. Kidron caught the attention of Americans from photographs showing the general astride the horse in a victory parade on the Champs-Élysées in Paris, and then in a parade through the Victory Arch in New York City at the end of World War I. Kidron died in Virginia in 1942 at about the age of thirty-five. He was easily identified by his white rear feet.

Kincsem (Horse)

Kincsem, whose name is Hungarian for "my precious" or "my treasure," was born in Kisber, Hungary, in 1874. Her breeder and owner, Erno Blaskovich, must not have imagined that this little Hungarian foal would grow up to race fifty-four times and win fifty-four times. Despite being a Thoroughbred, she was apparently quite an ugly duckling. She was described as having legs like stilts, a tail like an old broom, a skinny front and back, pot-bellied middle, and lop-sided ears. One night a man stole little Kincsem, and when her owner traced the thief to his camp he asked why he chose Kincsem to steal when he could have stolen one of his magnificent horses in his stables. The thief replied, "Because she will be the greatest of them all." Robert Hesp trained Kincsem and a year later she was entered in her first race in Budapest; she ran twelve lengths ahead of the field. In her second year of racing, she showed

less and less interest coming to the starting line, and in race after race she was left behind, yet she always managed to win. Kincsem died on March 17, 1887, at the age of fourteen and was mourned by royalty and commoner alike.

King Neptune (Pig)

King Neptune, a Hereford swine, was born on May 16, 1942, in a litter of 12 on the Sherman Boner farm in West Frankfort, Illinois. Sherman Boner's daughter Patty raised Neptune as a 4-H project, and then he was donated on December 5, 1942, to the United States Navy. If not for Navy recruiter Don C. Lingle, King Neptune would have been served at a fundraising pig roast. Instead, between 1942 and 1946, Lingle teamed up with auctioneer L. Oard Sitter, traveling through southern Illinois and auctioning the pig to raise war bonds to pay for the construction of the Iowa-class battleship *Illinois*. Neptune often made appearances wearing a navy-blue blanket, a crown and silver earrings. In 1945, a check was written to King Neptune, and Lingle obtained the pig's endorsement for the bank. Lingle and Neptune raised $19 million in war bonds (nearly $300 million in 2022 when adjusted for inflation). In 1946, Lingle once again saved Neptune—this time from the Chicago Stockyards—and placed him on a local farm for the rest of his life. On May 14, 1950, King Neptune died of pneumonia and was buried with military honors. Written on his tombstone were these words:

"King Neptune (1942–1950) [*sic*] Buried here—King Neptune, famous Navy mascot pig auctioned for $19,000,000.00 in war bonds 1942–1946 to help make a free world."

King Tut (Dog)

In 1917, while Herbert Hoover was organizing war relief in Belgium, he obtained a Belgian shepherd, King Tut. During his presidential campaign in 1928, a photograph of Hoover holding King Tut and smiling was credited with helping win the election. His public image was somewhat stiff and severe, and the photograph revealed a warmth that had previously eluded Hoover. *The New York Times* called it "one of the happiest pictures ever made" of Hoover. At the White House, King Tut fetched Hoover's newspapers and became a sentry, patrolling the grounds no matter the weather. The dog rarely slept and when Hoover tried to keep him indoors he barked until he was let out. King Tut was so stressed out that he quit eating and was sent to a quiet home in Virginia, but he did not improve. King Tut died in late 1929, at age eight, "worrying himself to death."

This photograph of Herbert Hoover holding King Tut was credited with playing a role in Hoover winning the presidential election in 1928 (photograph by Herbert E. French, Library of Congress).

Knut (Polar Bear)

Knut was born on December 5, 2006, at the Berlin Zoological Garden. Two cubs were born to his mother, Tosca, a 20-year-old former circus performer, and his father, Lars, a 13-year-old bear from the Tierpark Hellabrunn in Munich. For unknown reasons Tosca rejected her two cubs; therefore zookeepers had to rescue them from their enclosure. Knut's brother died four days later, and Knut became the first polar bear in 30 years to be born and survive in the Berlin Zoo. After Knut spent his first 44 days in an incubator, zookeeper Thomas Dörflein became his caretaker. Dörflein filled Knut's need for round-the-clock care by sleeping on a mattress beside his crate; he played with, bathed and fed the cub daily and also accompanied Knut on his twice-daily public shows. On March 23, 2007, Knut appeared to the public for the first time, and the occasion was dubbed "Knut Day" to commemorate his first appearance worldwide. At seven months old and weighing 110 pounds, Knut became a safety concern. In July 2007, the zoo cancelled his public shows. Zoo spokeswoman Regina Damm said it was time for the bear to "associate with other bears and not with other people." In November 2007, with Knut weighing over 198 pounds, he was thought to be dangerous and his interaction with human handlers ceased. Knut's first birthday was celebrated by hundreds of children and broadcast on German television, and the national mint issued 25,000 commemorative silver coins. Dörflein died of a heart attack at the age of 44 on September 22, 2008. Knut died on March 19, 2011, in his enclosure at age four. Six hundred to 700 visitors witnessed Knut's death, reporting that his rear left leg began shaking and then he began convulsing and fell backwards into the pool. A pathology report indicated that the immediate cause of death was drowning. In January 2014, the *Journal of Comparative Pathology* and the Leibniz Institute for Zoo and Wildlife Research published Knut's full post-mortem autopsy results. They led experts to hypothesize that he had had a virus that caused encephalitis and that his brain was so severely damaged that he would have died anyway. In August 2015, his official cause of death was reported to be from anti-NMDA receptor encephalitis. On October 24, 2012, the Berlin Zoo unveiled Ukrainian artist Josef Tabachnyk's bronze sculpture *Knut-The Dreamer,* showing the bear "stretching out dreamily on a rock." There was also a traveling exhibit of a full-sized sculpture covered in Knut's fur. Knut was a media sensation, spawning many commercial successes including but not limited to: toys, "Knut" candy, songs, books, DVDs, and a film. Knut was also championed for environmental causes.

Kola (Dog)

Kola's owners, Mr. and Mrs. William Thress, thought that they had a normal, loving dog with some husky in him. What they did not expect was that their beloved dog would become a hero. The day started when six-year-old Ronnie Sowell decided to sneak out of his yard to explore more of his world. That afternoon a vicious storm broke out in Dana Point, California, and as time went on and the storm got worse the town became frantic about missing Ronnie. Kola knew Ronnie and had seen him walk away. After searching all through the night the posse found Kola lying on top of Ronnie, keeping him completely dry and safe. It was puzzling because Kola had no training in tracking. On April 25, 1959, Kola's heroism was recognized with an award from the Los Angeles branch of the Society for Prevention of Cruelty to Animals.

Kostya (Dog)

Kostya, a German shepherd from Tolyatti, Russia, was the sole survivor of a car accident. After the accident, people began to notice the dog standing and looking, presumably waiting for his deceased owner's return. People built him dog houses and, try as they did to adopt him, Kostya refused to leave his vigil. After seven years, in 2002, Kostya was found dead in the woods. Touched by the story, citizens installed a billboard with the saying, "Dog, teach us love and devotion." As of 2003, on the spot where Kostya stood vigil, a bronze sculpture sits with Kostya's head looking east.

Kudryavka *see* Laika

Kurt (Dog)

Kurt, a Doberman, was the first dog killed during the Battle of Guam. His statue sits atop the "Always Faithful" Military Working Dog Monument at the Marine Corps Heritage Center in Triangle, Virginia. On one side of Kurt's collar is the word ALWAYS and on the other side is FAITHFUL. The monument honors the twenty-five war dogs who died in service during the Battle of Guam in 1944. Kurt saved many lives when he silently warned of an incoming attack of Japanese soldiers. Kurt and his handler were both seriously wounded by mortar shell. The soldier refused treatment until Kurt was brought to the rear. Kurt died in his arms, too wounded to be helped.

Laddie Boy (Dog)

President Warren G. Harding's first gift as president was Laddie Boy, an Airedale puppy, in 1921. Harding adored him from the start and the two became constant companions. Laddie Boy delivered the morning paper, fetched lost

Laddie Boy, shown here in July 1922 in a photo by the National Photo Company, was the first gift that Warren G. Harding received as President (Library of Congress, LC-USZ62-131900).

balls on the golf course, and was trained to sit in his own hand-crafted chair at Cabinet meetings. Laddie Boy was a regular feature at fundraising events that the first lady attended and became a poster child for the national promotion of animal rights. Harding commissioned one thousand miniatures of Laddie Boy to hand out to his supporters. Today, historians talk about Laddie Boy as the most celebrated of all presidential dogs. The White House held birthday parties for him and other dogs were invited. It is said that for three days prior to Harding's death, Laddie Boy was restless and howled inconsolably. After Harding's death, Laddie Boy spent his remaining years with Harry Barker, a U.S. Secret Service agent, and lived a good life in Boston with the Barker family. Laddie Boy died in 1929. As a tribute to Harding, newsboys collected 19,134 pennies to be melted and sculpted into a statue of Laddie Boy, which was completed in 1927. Unfortunately, Harding's widow died before it was completed. The statue was presented to the Smithsonian Institution.

Lady see **Titanic Animals**

Lady Astor (Pigeon)

Lady Astor served on the front lines in North Africa during World War II. She flew a sixty-mile flight with an urgent message. During her flight she was shot by pellets, broke a leg and lost feathers. Once home after delivering her message she immediately collapsed. After being nursed back to health, she was retired from service.

Lady Jane (Pigeon)

Lady Jane was gassed and still managed to fly back to her loft. Reports show that she recovered from her injuries.

Lady Wonder (Horse)

In Richmond, Virginia, Claudia Fonda owned a horse named Lady Wonder, who was thought to have psychic powers. Dr. J. B. Rhine and Dr. Louisa E. Rhine of Duke University wrote a paper on their investigation of Lady Wonder. Researchers at the Department of Psychology at Duke University concluded that Lady Wonder was capable of reading a person's mind. Thanks to their research, Lady Wonder became the most famous non-racing horse in the world.

Some of Lady Wonder's alleged abilities:

- A spectator took a coin from his pocket. No one else saw the face of it. What was the date on it? The mare nosed over the blocks, "1–9–1–4." Correct.
- "Who in the group has on a pink dress?" the mare was asked. "As I live!" exclaimed the woman in pink as the mare thrust her head emphatically in the visitor's direction.
- A spectator picked up a clock and turned the dial to ten minutes past six. Nobody else saw the time, and he thrust the clock face against his body. "What time is it by this clock?" the mare was asked. "Six-one-naught," replied Lady.
- "What is the sum of eight and seven?" asked a spectator. And the mare answered lackadaisically, "Fifteen."

- A visitor held a pocket knife in his hand. He inquired, "What have I here?" And the mare spelled out "k-n-i-f-e."
- "What is the name of this boy at my side?" was the question. And the mare replied, "Leroy."

Laika (dog)

History remembers Yuri Gagarin as the first man in space in 1961, but he was preceded in 1957 by Laika (originally named Kudryavka), a female terrier mix who was scooped up from the streets of Moscow and rocketed into history—becoming the first animal to orbit the Earth. Launched in the Russian satellite *Sputnik* 2, Laika became a global celebrity as the world marveled at her bravery. Awe quickly turned to anger as it was revealed not long after the launch that the Soviets did not have a way to return Laika safely to Earth, which stirred up a lot of discussion on the ethics of the flight. She likely died within hours after launch due to overheating, and her remains were incinerated as *Sputnik* 2 re-entered Earth's atmosphere. The experiment did provide scientists with data on the effect of spaceflight on humans. Some Russian scientists today regret the incident because the space industry probably did not learn enough to justify Laika's horrible

Laika was the first animal to orbit the earth aboard the Russian satellite *Sputnik* 2.

death. On April 11, 2008, Russian officials unveiled a monument to Laika featuring a dog standing on top of a rocket.

Yergeniy Shabarov remembered:

> After placing Laika in the container and before closing the hatch, we kissed her nose and wished her bon voyage, knowing that she would not survive the flight.

Lancelot (Goat/Unicorn)

Lancelot, a living unicorn? Lancelot was a goat turned into a fictional unicorn, having a large horn protruding from the middle of his head. The press were shown an X-ray by Dr. Charles Reid, professor of radiology at the University of Pennsylvania, showing that the horn was indeed part of the goat's skull, not an implant or artificial addition. On his visit to New York in April 1985, Lancelot created quite a controversy among animal rights groups; however Dr. William Donawick declared him "content, healthy and living." Therefore, he was validated to be a "content, healthy, living unicorn." When the Ringling Bros. and Barnum & Bailey Circus came to Houston in July 1984, their spokespeople claimed that Lancelot had just "wandered up to a tent." Enchanted by this animal, they made him part of their shows around the country; he was cared for by Heather Harris, a former dancer. In early 1985, the circus trademarked the name "Living Unicorn," which they used in the advertisements of their shows. In April 1985, when the show came to New York, both The Humane Society and the American Society for the Prevention of Cruelty to Animals (ASPCA) became concerned over how Lancelot was promoted to be something out of a storybook. The original owner of Lancelot, Oberon Zell, a self-professed wizard, read Peter S. Beagle's novel *The Last Unicorn* and began studying the mention of the unicorn through history, which led him to the work of Franklin

Dove, a biologist in the 1930s. Dove had discovered a surgical procedure for fusing the horns of a goat together. One would begin by getting a goat less than a week old whose horns (buds) were still part of the skin and not yet connected to the skull, maneuver them close together, and stitch them to meet in the center of the forehead; as the goat grew the horns would merge. Zell used Dove's notes in his own work in 1980. He cross-bred Angora goats with Saanen goats and was successful in growing a single horn. USDA chief veterinarian Dr. Gerald Toms confirmed Lancelot to be a goat that had had a grafting procedure. In a press conference, Ringling Brothers showed X-rays confirming the horn's natural growth.

Larry (Cat)

Larry is known for being the 10 Downing Street cat and the UK Cabinet Office's Chief Mouser. Larry was a brown and white stray tabby cat rescued by the Downing Street staff from the Battersea Dogs and Cats Home in 2011. Originally meant to be a pet for Prime Minister David Cameron and his family, Larry took up the duties of Sybil, the former Downing Street cat that died in 2009. Larry's official duties per The Downing Street website included "greeting guests to the house, inspecting security defenses, and testing antique furniture for napping quality." It was revealed later in 2011 that Larry was sleeping more than hunting and keeping company with Maisie, a female cat. From 2012 to 2014, Larry shared his role of Chief Mouser with George Osborne's cat Freya. The Battersea Dogs and Cats Home honored Larry with a blue plaque in October 2012.

Lassie *see* **Pal**

Lazarus *see* **Bummer and Lazarus**

Lestock (Coyote)

In 1915, on their journey across Canada and ultimately on to Europe, the Loyal Edmonton Regiment, 49th Battalion, stopped in Lestock, Saskatchewan. A resident gave the men a coyote pup. The battalion adopted him as their mascot and named him after the town. Lestock was given to a zoo in London before the 49th Battalion went into battle. Today Lestock's face appears in the center of the unit's badge.

Lexington (Horse)

Lexington was a United States Thoroughbred racehorse born on March 17, 1850. Racing at ages three and four, he won six of seven races and finished second once. On April 2, 1855, Lexington set a record running four miles in 7 minutes, 19 and three-quarters seconds at the Metairie race course in New Orleans. Over his racing career his total earnings were $56,600. Lexington had to retire in 1855 due to failing eyesight. After his retirement he became the leading sire in North America; he sired 236 winners that won 1,176 races, ran second 348 times and third 42 times— totaling $1,159,321 in prize money. Lexington died on July 1, 1875, and in 1878 his bones were donated to the U.S. National Museum (now the Smithsonian Institution). For many years his skeleton was on exhibition in the Osteology Hall of the National Museum of American History. In 1999 he was part of the museum's exhibition titled "On Time."

Lila (Dog)

Lila, a fox terrier belonging to British General William Howe, ended up revealing the character of George Washington. In July 1777, General Howe moved his men with the intention of seizing the city of Philadelphia, which was serving as the revolutionary capital. American spirits were low after some serious defeats in September 1777, which gave Cornwallis the momentum to march into Philadelphia and claim it for the British. General Howe's next move was to send his army to Georgetown. With the British spreading out, Washington made a plan to overtake the forces at the garrison in Germantown. His men's lack of preparedness and foggy weather on the battlefield assured that for the remainder of the war Philadelphia would remain under British rule. It is at this point that Lila became a part of history. The Americans captured a small dog found on the battlefield, and from her collar they knew she belonged to General Howe. Despite his men wanting to hold the dog in retribution, Washington showed his true character by sending a messenger to return the dog along with a note reading: "General Washington's compliments to General Howe, does himself the pleasure to return [to] him a Dog, which accidentally fell into his hands, and by the inscription on the Collar appears to belong to General Howe." This documented note, written in the handwriting of Washington's aide-de-camp Alexander Hamilton, can be found in the *George Washington Papers* collection at the Library of Congress. In most sources referring to General William Howe's dog, there is no name of the dog mentioned.

Lin Wang (Elephant)

Lin Wang was an Asian elephant serving during the Second Sino-Japanese War, which began in 1937. However after the attack on Pearl Harbor in 1941, the war began to play a much greater part in World War II. In 1942, the Japanese invaded Burma and used work elephants to build roads and transport supplies. A year later Chinese Expeditionary Forces captured 13 of their elephants, marching them to

China. At this time, Lin Wang was named Ah Mei, meaning "the beautiful." Six of the elephants died and the war had ended upon their arrival in Guangdong. After the war, the remaining seven elephants were still active in building war monuments and raising money. In 1946, they were used to raise money for famine relief by performing in a circus. Then four of the elephants were given to four different zoos. Lin Wang and the two other remaining elephants traveled with Sun Li-jen to Taiwan. One elephant got sick and died and the other two were used for simple tasks around the army base in Fongshan, Kaohsiung. The one remaining elephant, Lin Wang ("forest king"), was donated to the Taipei Zoo in 1954. He became a popular and beloved attraction until his death at age 86 in 2003.

Ling-Ling and Hsing-Hsing (Giant Pandas)

These two giant pandas were captured in the wild in 1971—Ling-Ling in June and Hsing-Hsing in December. Following President Richard Nixon's 1972 visit, the Chinese government gifted the pandas to the United States. On April 16, 1972, they arrived at the National Zoo in Washington, D.C., and on April 20 the pandas were formally received at a ceremony attended by First Lady Pat Nixon. In December 1992, Ling-Ling died from heart failure, and in November 1999 at age 28, Hsing-Hsing was euthanized due to kidney failure. While at the zoo the pair had five cubs; however they only survived a few days. A year after Hsing-Hsing died the pandas were replaced by Mei Xiang and Tian Tian from the China Research & Conservation Center for the Giant Panda.

Little Diamond (Elephant)

Born on March 2, 1978, at the Knoxville Zoological Park, Little Diamond was the first African elephant born in captivity in North America. She stood less than three feet tall and weighed 200 pounds. Her mother, Toto, was on breeding loan from the Bronx Zoo in New York. Her father, Diamond, had been donated to the zoo in 1960 by the Ringling Bros. and Barnum & Bailey Circus. Little Diamond joined the herd at the North Carolina Zoological Park near Asheboro in 1995. Little Diamond died in 2014 of a stomach ailment—an impacted large intestine from ingesting sand. Sand is a common material used in zoo elephant holding facilities and exhibits.

Little Fool (Starling)

There are records that for three years Wolfgang Amadeus Mozart kept a pet starling he had purchased from a bird store. Mozart admired the starling's ability to mimic new sounds, even Mozart's melodies. It is impossible to confirm, but some have claimed that Mozart even wrote some of his concertos from inspiration taken from the bird's songs. What has been confirmed is that upon Little Fool's death Mozart grieved deeply and organized a funeral. He even composed an epitaph, making it clear how he genuinely loved his pet bird: "Here lies a little fool, whom I held dear; a starling in the prime of his brief time, whose doom it was to drain death's bitter pain. Thinking of this, my heart is riven apart!"

Little Jim (Dog)

Little Jim was a famous British messenger dog of an extremely strange mix of retriever and Pomeranian. The soldiers were amazed at how fast he could run, being that he was a small dog. He was never gassed or shot during his days as a messenger dog.

Little Jim (Dog)

It was common during World War I for several dogs to have the same name. This Little Jim ("Dog 36") was a cross between a retriever and a spaniel. He was deemed the finest dog in France by the commanding officer. Little Jim's handler, Private Jim Osbourne, cited one occasion of Jim's quickness in delivering messages when under heavy fire during an offensive in Belgium. He also cited one time in the trenches when Jim warned the soldiers of gas attacks and arrived at headquarters in fifteen minutes with the message about the gas before a wire came through.

Little John (Coyote)

Artist Joseph Beuys' past experiences on how he was burned and became an artist are in doubt. He was an art teacher, sculptor and performer. In 1974, he arrived at Kennedy Airport to begin his most famous performance piece. A vintage emergency vehicle sped Beuys, wrapped in a rug, to a downtown gallery. He was dumped in a corner that was furnished with a pile of straw, a wooden cane, a flashlight, fifty copies of *The Wall Street Journal* and a live coyote. In the piece, called *Action*, he remained in the cage with the coyote for a week. He named the coyote Little John. Footage and photographs show Little John alert and lively, with a charismatic face. Little John probably came from a ranch in New Jersey. After a week, *Action* ended and Beuys left without Little John. Elena Passarello wrote about Little John's mythical reputation, which he deserved. There is no confirmed information about Little John.

Little Sorrel (Horse)

Also known as "Old Sorrel," Little Sorrel was General Stonewall Jackson's beloved horse. In 1861, the general's troops seized a train carrying a carload of horses at Harper's Ferry. Jackson chose a small Gelding and named it "Little Sorrel." For the next two years, Jackson rode Little Sorrel almost exclusively. On May 2, 1863, the troops accidentally shot Jackson and eight days later he died. Little Sorrel was sent to live with Jackson's family. Outliving the general and to a very old age, he toured county fairs and attended Confederate functions. The hair from his mane and tail was often pulled out by visitors for souvenirs. Little Sorrel died in 1886 and his bones were buried by the Jackson statue at the Virginia Military Institute in Lexington, Virginia. His stuffed hide was on display in Lexington in the VMI Museum until 1997, when he was interred.

Little Yellow Jacket (Bull)

Little Yellow Jacket was the most successful riding bull in the history of the Professional Bull Riders organization.

Lloyd (Dog)

Handler Errington tells of how his three dogs, Jack the Airedale and two large Welsh terriers, Whitefoot and Lloyd, worked in Strazeele (in northern France) through difficult roads littered with the remnants of shelling.

Lobo (Wolf)

Lobo, "King of Currumpaw," led a pack of wolves that attacked livestock in northeast New Mexico for five years in the 1890s. Ernest Thompson Seton reported in *Wild Animals I Have Known* (1898) how he was invited to catch Lobo after several hunters had failed. He tried poison and buried traps, but Lobo's undoing was when he witnessed the death of his mate, Blanca (a white wolf), that was killed by Seton. Lobo howled for two days straight. Touched by Lobo's loyalty to his mate, Seton could not kill him. He used Blanca's body to securely trap Lobo and humanely chained him in a pasture with meat and water nearby. Lobo just lay there and the next morning he was found dead.

Lonesome George (Tortoise)

Lonesome George, a male Pinta Island tortoise (Chelonoidis abingdonii), is aptly named because until he was discovered in November 1, 1971, it was believed his species was extinct. Discovered on Pinta Island by Hungarian malacologist Jozsef Vagvolgyi, Lonesome George belonged to a subspecies of Galapagos tortoise that was native to Ecuador's Pinta Island. In 1877, herpetologist Albert Gunther was the first to describe the species. By 1900, the natural habitat of the Pinta Island giant tortoises had been destroyed by hunters and goats. George was relocated to the Charles Darwin Research Station on Santa Cruz Island and cared for by Fausto Llerena. All attempts to produce offspring were unsuccessful, but not due to lack of trying. In July 2008, a female mate laid 13 eggs, but by December 2008 all the eggs were inviable. On July 23, 2009, another female mate laid 5 eggs, but on December 16, after their incubation period, the eggs were also inviable. This Pinta tortoise became one of the rarest creatures in the world. Lonesome George was found dead on June 24, 2012, by his caretaker, and the cause of death was most likely from heart failure. His frozen body was shipped to the American Museum of Natural History in New York, where he was preserved by taxidermist George Dante. After some disputes as to where Lonesome George should be housed, on February 17, 2017, Lonesome George was returned to the Galapagos Islands, where an entire exhibit dedicated to him is at the Charles Darwin Research Station. The November 2012 *Biological Conservation* reported that researchers had identified 17 tortoises that were partially descended from the same subspecies as Lonesome George. In December 2015, Yale University researchers discovered a species, Chelonoidis donfaustoi, with a 90 percent DNA match to the Pinta tortoise. Scientists speculate that perhaps this discovery might be used in resurrecting the species. Lonesome George will be remembered for contributions to animal conservation efforts. At present, approximately 20,000 subspecies of the giant tortoise live on the Galapagos.

Lord Adelaide (Pigeon)

Lord Adelaide was a famous tank pigeon who made two successful flights through flying bullets without being shot. On his third run he was not so lucky and was shot. Despite being bloody and weak, he made it back to his loft.

Lord Fauntleroy (Mule)

In 1976, the United States was celebrating the bicentennial and people were feeling especially patriotic. Chuck Waggoner and Randy Scheiding created the Great American Horse Race, which would cover 3,500 miles over 14 weeks, encompassing parts of the Oregon Trail, the Pony Express Trail, and the Donner Party's journey and offering a prize of $25,000. On May 31, 1976, Arabian stallions, Icelandic horses, tall Irish Thoroughbreds, Appaloosas, and Virl Norton's Lord Fauntleroy "Leroy" the mule turned up in Frankfort, New York, ready to start the race. Ninety-one teams lined up at the starting line: the youngest competitor was eighteen, the oldest sixty-nine, and among the riders were chiropractors, pediatricians, grad students, nurses, farmers, cowboys and one university president. In the end, Leroy won with 315.47 hours under the saddle, and in second place was an Arabian with 324.6 hours with penalties. Toledo's *The Blade* wrote, "Mule Runs Away with Great American Horse Race" and "Lord Fauntleroy Is No. 1." Norton referred to himself as the "Great American Horseman," and Lord Fauntleroy as the "Great American Horse."

Lulu (Kangaroo)

In September 2003, during a storm in Victoria State, a branch fell, hitting farmer Len Richards and knocking him unconscious. Lulu, his pet kangaroo, skipped to the house and alerted Len's wife Lynne by "barking" at her. While searching for her husband, Lynne and her nephew found Lulu standing over Len. When he recovered, he told Australian radio, "My nephew, when he got to my side, said she'd actually tipped me on my side and vomit was coming out of my mouth so she'd actually saved me from choking." Lulu was awarded the National Animal Valor Award.

Lulu (Vietnamese pot-bellied pig)

In 1997, JoAnn and Jack Altsman agreed to babysit their daughter's pig, Lulu, and it turned out to be a permanent new home for Lulu. In Pennsylvania in 1998, JoAnn Altsman suffered a heart attack and passed out alone except for Lulu and Bear, their dog. Her husband was on a fishing trip. Lulu forced herself outside and into the street, causing the animal to bleed. She was ignored until she lay in the street playing dead. A man stopped his car and followed Lulu inside to JoAnn. Doctors believe that if help had arrived just fifteen minutes later, JoAnn would not have survived. After saving JoAnn's life, Lulu became a media celebrity, even appearing on *Good Morning America* and *The Oprah Winfrey Show*. Lulu received the Tiffany Gold Hero's Medal from the American Society for the Prevention of Cruelty to Animals.

Lummo (Cat)

From 1934 to 1937, the schooner *Penola* carried out the British Graham Land Expedition. The mission of the expedition was to survey the west coast of Graham Land to gain knowledge of the whole region and explore the passage through the Stefansson Strait to the Weddell Sea. Lummo, a white cat with patches of black, was presented to the crew by the Rev. Harold E. Lumsdale, dean of Stanley Cathedral, while the ship was stocking up on provisions in the Falkland Islands. Lummo seemed to adapt to the cold of the Antarctic and was known to curl up in the snow. Lummo was one of several cats that left with the expedition in 1934, but he is the only one to have survived. So popular was Lummo with the men that by general consent he became "the First Marquess of Lumsdale" for "gallant bearing in adversity." Perhaps it is not surprising that the pet thrived, for his diet would have filled poor home-loving cats with envy. He lived on the fat of the land, seal's liver and a breast of penguin for dinner. He never tired of this menu. Lummo survived the three years of the expedition and afterwards went to live with a crew member's family in Woking, Surrey, until the cat's death during World Word II.

In August 1937, the *Auckland Star* ran a piece about Lummo:

No member of the Graham Land Antarctic expedition, recently returned to England, has attracted more public attention than Lummo. Lummo is the expedition's cat, hailed as the only cat which has ever survived the Polar climate. At the age of three-and-a-bit, he can boast of having shared all the rigors of Antarctic exploration for more than two years.

Lump (Dog)

Pablo Picasso acquired Lump, a dachshund, in 1957 from a friend. It was true love at first-sight when they met. Lump was allowed anywhere in Picasso's studio and appeared in fifty-four of his works of art. They were together for sixteen years and died within months of each other. Lump died on March 29, 1973.

Lutz (Dog)

Lutz was a famous dog awarded the War Cross Star for advancing and alerting the troops to the presence of the enemy.

Macaroni (Horse)

Macaroni, a piebald pony, was a gift from President Lyndon B. Johnson to President John F. Kennedy's daughter Caroline. This pony inspired a New York composer to write a song titled "My Pony Macaroni." Macaroni was also the center of attention for a group called The Society for Indecency to Naked Animals, which was created by Alan Abel and Buck Henry, respectively the group's spokesperson and president. Naked animals were destroying the moral integrity of our great nation! This hoax—a commentary on conservative moralism—gained media attention and actual members. Abel was the master of the hoax, tricking newspapers and many others. Later, Neil Diamond disclosed that a photo of Caroline and Macaroni had inspired the song "Sweet Caroline."

Macavity *see* **Commuting Animals of England**

Macaroni pictured here with President John F. Kennedy, his son John Jr. and his daughter Caroline, June 22, 1962 (Robert Knudsen. White House Photographs. John F. Kennedy Presidential Library and Museum, Boston).

Macek (Cat)

Nikola Tesla's most famous contribution to science was the creation of the alternating-current (AC) electric system, which is still used in the world today. His discovery was made possible by his favorite childhood companion, a black cat named Macek, who sparked his lifelong interest in electricity. On a podcast episode about cats, Will Pearson read part of a 1939 letter Tesla wrote to a curious 12-year-old of how his discovery came to be: "In the dusk of the evening as I stroked Macek's back, I saw a miracle which made me speechless with amazement. Macek's back was a sheet of light, and my hand produced a shower of crackling sparks loud enough to be heard all over the house.... My mother seemed charmed—'Stop playing with the cat,' she said. 'He might start a fire.' But I was thinking abstractly. Is Nature a gigantic cat? If so, who strokes its back? ... I cannot exaggerate the effect of this marvelous night on my childish imagination. Day after day I have asked myself, what is electricity?"

Madame Theophile (Cat)

The French poet and writer Theophile Gautier (1811–1872) truly adored cats. Many of his poems are about them, and they were his beloved companions. In his book titled *La Menagerie Intime*, or *My Private Menagerie*, he wrote about his household pets, which was summarized in an 1895 article by *The Daily Telegraph*: "Madame Theophile was another favorite, reddish and white breasted, pink-nosed and blue-eyed. She dwelt with him on terms of great intimacy, sleeping with him, sitting on the arm of his chair when he wrote, following him on his walks through the garden and always present at meals, when she sometimes stole attractive bits from his plate. She had all the tastes of a great French lady.... She liked music but sharp, high notes affected her and she would put her paw upon a singer's lips when such a high note distressed her."

Gautier told an amusing story about Madame Theophile and her encounter with a parrot that was left in his charge for a short time. Madame had never seen a parrot and she sat staring at it, trying to figure it out. The parrot started to become alarmed and ruffled its feathers. The cat started to make a move and the parrot said in a deep voice, "Have you breakfasted, Jacquot?" Madame became scared and more so when the parrot spoke again. Madame, fleeing for her life, went under the bed.

Maharajah (Elephant)

The historical book *The Elephant Thief* by Jane Kerr is based on the true story of this elephant. In 1872 Maharajah was sold at auction in Edinburgh to the owner of Belle Vue Zoological Gardens in Manchester, England. After being sold, he, with his keeper, walked from Edinburgh to Manchester in 10 days. Maharajah's bones lie in the Manchester Museum.

Man o' War (Horse)

Man o' War, an American Thoroughbred, is widely considered one of the greatest racehorses of all time. During his career just after World War I, he won 20 of 21 races, earning $249,465 in prize money. Man o' War was a chestnut horse with a white star and stripe on his forehead and was foaled on March 29, 1917. Man o' War was the odds-on favorite in every start of his career and typically won in front-running fashion. Developing his talent was not an easy task because of his occasional wild temperament. Man o' War was retired as a three-year-old after his undefeated season and became a stud. Man o' War sired the famous racehorse Hard Tack that sired Seabiscuit, a symbol of hope during the Great Depression. Man o' War was retired from stud in 1943 after suffering a heart attack. He died at age thirty of another apparent heart attack.

An editorial in the November 3, 1947, edition of *The New York Times* stated:

> No other horse ever won such fame as Man o' War. None was more beautiful, with lovelier lines of grace and power. None was more beloved by an admiring and faithful public. Few have lived so long. The American scene seems a little vacant with Man o' War gone to the Elysian Fields where all good horses go.

Mancs (dog)

A famous rescue dog from Hungary, Mancs (whose name means "paw") was a member of the Spider Special Rescue Team of Miskolc, Hungary. Mancs and the team traveled around the world to search for survivors after earthquakes. Mancs was known for his keen sense of smell and the clear signal he sent to rescuers to indicate if someone was alive under the rubble. A statue of Mancs was erected in Miskolc in 2004, two years before his death.

Marengo (Horse)

Marengo, a grey Arabian, was the famous war horse of Napoleon I of France and named after the Battle of Marengo. In 1799, at six years old Marengo was imported to France from Egypt. The war horse carried the emperor in the Battle of Austerlitz, Battle of Jena-Auerstedt, Battle of Wagram, and Battle of Waterloo, and was often used in 80-mile gallops from Valladolid to Burgos. Marengo fled with 52 of Napoleon's horses in a raid by the Russians in 1812. In 1815, he was captured by William Henry Francis Petre, 11th Baron Petre, at the Battle of Waterloo. Petre sold him to Lieutenant Colonel Angerstein of the Grenadier Guards. Marengo died at age 38 and his skeleton (minus two hooves) was preserved and displayed at the National Army Museum in Chelsea, London. One of the hooves was given to the officers of the Brigade of Guards, as a snuff box with a silver lid inscribed, "Hoof of Marengo, barb charger of Napoleon, ridden by him at Marengo, Austerlitz, Jena, Wagram, in the Russian campaign, and at Waterloo"; the other hoof was mounted as a silver inkwell remaining with the family. During his career Marengo was wounded eight times.

Marian (Grizzly Bear)

Marian (AKA Bear Number 40) was two-and-a-half years old in 1960 when she was captured, sedated, and tagged on the ear. In September 1961, she became the first grizzly to be tracked by radio collar as part of the long-range study of grizzly bears in Yellowstone National Park. During the next eight years, Marian's mating, preparing for winter dens, and raising of several families were observed. In the spring of 1964, Marian emerged from hibernation with her first two cubs. A surviving cub weighed over 100 pounds by September, and Marian weaned him the following June. In 1966, Marian was seen nursing and mothering a second pair of cubs. In October 1969, a ranger trying to tranquilize one of her yearlings in order to move it shot and killed Marian when she tried to protect her cub.

Marjan (Lion)

In Persian, Marjan means "coral." Marjan was born in 1976 and in 1978 the Cologne Zoo gave him to the Kabul Zoo in Afghanistan. Shortly after his arrival in Kabul, Marjan was joined by a lioness named Chucha. Over the next 23 years, he survived many conflicts in the country. At the zoo, he lived through the arrival of the communist People's Democratic Party of Afghanistan, the Great Saur Revolution, and the USSR invasion. In 1989, a civil war broke out after the Russians left the country. In all the chaos the zoo had been shelled many times and fell into disrepair. Tragedy befell Marjan on March 27, 1995: On a bet, a man entered his cage and was able to stroke Chucha, but when he tried to touch Marjan the lion killed the man. The next day the man's brother threw three hand grenades into Marjan's den, which left the lion blind, deaf, and permanently disabled. A week later the man was attacked and died. The arrival of a United States–led coalition sparked the media's interest, drawing many animal welfare organizations, such as World Animal Protection, and Association of Zoos and Aquariums, to aid with money, medical supplies, vets and nurses. On January 25, 2002, Marjan died of old age. Public and private funeral ceremonies were held and he was buried in the zoo. Posted on his grave in Pashto are these words: "Here lies Marjan, who was about 23. He was the most famous lion in the world." In March 2002, China donated a pair of lions to the Kabul Zoo.

Marjan is either mentioned or featured in several different works of fiction:

Khaled Hosseini's novel *The Kite Runner*
Marjan is the central figure of the play
    The Lion of Kabul by Colin Teevan
Part of *The Great Game: Afghanistan*
Marjan appears in Denis Johnson's essay
    collection *Seek*
Marjan is mentioned in Kim Barker's
    book *The Taliban Shuffle*
A deleted scene in *Whiskey Tango
    Foxtrot* features the attack on Marjan
Marjan is the subject of Dwyer Jones'
    poem "Lion of Afghanistan (Apology
    to Marjan)"

Marley (Dog)

A *New York Times* bestselling autobiographical book by journalist John Grogan, *Marley & Me: Life and Love with the World's Worst Dog*, published in 2005, reflects on the thirteen years his family lived with Marley, their yellow Labrador retriever. Marley was not very well-behaved and was described as being destructive, high-strung, boisterous, strong, powerful, always hungry, very energetic, and somewhat dysfunctional. Despite his uncontrollable behaviors, all is forgiven as the family realizes he has a "heart of gold," and seeing the devotion, love and trust he shows towards the family, they grow to accept him as he is. Their grief when Marley dies from gastric dilatation volvulus (a stomach torsion condition) in his old age forms the background for the story.

The story *Marley & Me* has spawned the writing of three additional books:

Marley: A Dog Like No Other, adapted
    for younger readers
Bad Dog, Marley!, a picture book
A Very Marley Christmas

On December 25, 2008, a family comedy-drama film, *Marley & Me*, was released. A prequel followed in 2011, *Marley & Me: The Puppy Years*. In an interview Grogan concluded, "Commitment matters. That 'in good times and bad, in sickness and in health' really means something. We didn't give up on Marley when it would have been easy to, and in the end he came through and proved himself a great and memorable pet." (BookBrowse.com)

Marocco (Horse)

Marocco (AKA Bankes' Horse, after his trainer William Bankes) was named after the Moroccan leather used to make saddles. Bankes' educated horse was so famous in Elizabethan England that Shakespeare, Ben Johnson and Sir Walter Raleigh were impressed by Marocco's feats. The horse with silver horse shoes could dance, count, recognize colors, and read dice. In France, monks and friars accused Bankes of being a conjurer. Bankes then asked one of them to hold up a crucifix, and when Marocco knelt before it, all was forgiven. In 1600, Marocco climbed over a thousand steps to the top of Old St. Paul's Cathedral. The horse was sometimes referred to as the "Dancing Horse," the "Thinking Horse," or the "Politic Horse." He died at about age 20 in 1606, but had continued to tour Europe as late as 1605. After Marocco died, Bankes revealed the secrets of his training.

Martha (Pigeon)

Martha was the last known living passenger pigeon (Ectopistes migratorius); she was named Martha after First Lady Martha Washington. Arlie William Schorger, in his monograph, said of Martha that it was "difficult to find a more garbled history." There are several different versions of Martha's story, but it is generally accepted that by the turn of the 20th century, Professor Charles Otis Whitman at the University of Chicago had been acquiring his passenger pigeons from David Whittaker of Wisconsin. Whitman kept these pigeons, along with his other birds— rock doves and Eurasian collared doves— to study their behavior. Unsuccessful attempts at breeding these birds prompted Whitman to send Martha to the Cincinnati Zoo in 1902. Two other sources claim that Martha was a descendant of three pairs of passenger pigeons purchased by the Cincinnati Zoo in 1877. Another belief is that when the Cincinnati Zoo opened in 1885, there were 22 birds in its collection, and Martha was hatched in 1885. By November 1907, Martha and two companion male pigeons at the zoo were the only known surviving passenger pigeons. Four captive males in Milwaukee had perished during the winter. Then, one of the zoo's males died in April 1909, and the other one died on July 10, 1910, making Martha a celebrity as an endling—with a reward of $1,000 for finding a mate for Martha. Martha suffered an apoplectic stroke and a few years later she died, on September 1, 1914, of old age. Based on differing sources Martha was anywhere from 17–29 years old

Named after First Lady Martha Washington, Martha, pictured here in 1912, was the last known living passenger pigeon.

when she died, but it is generally thought that she was 29. After her death Martha was exhibited in many locations. William Palmer skinned her and Nelson R. Wood mounted her skin, and her internal parts were preserved and stored at the National Museum of Natural History. She was on exhibit at the museum's Bird Hall from the 1920s through the 1950s, sitting on a branch with another male passenger pigeon that had been shot in 1873. From 1956–1999, she was part of the Smithsonian's Birds of the World exhibit. However, in 1966, she was borrowed by the Zoological Society of San Diego's Golden Jubilee Conservation Conference as a mascot for conservation, and in June 1974 she went to the Cincinnati Zoo for a dedication of the Passenger Pigeon Memorial. Martha became a special exhibit at the Cincinnati Zoo after the Smithsonian shut down its Birds of the World exhibit. From June 2014 to September 2015 Martha was back at the Smithsonian for its "Once There Were Billions" exhibit. A Harvard historian wrote that Martha's remains were "an organic monument, biologically continuous with the living bird she commemorates, the embodiment of extinction itself." John Herald, a bluegrass singer, wrote a song titled "Martha (Last of the Passenger Pigeons)."

Mary (Elephant)

Mary was a five-ton Asian elephant performing in the Sparks World Famous Shows circus. Mary's demise really began in September 1916 in Kingsport, Tennessee, when she was accosted by new assistant handler Walter "Red" Eldridge. Mary responded by picking him up with her trunk, flinging him into a snack stand and then crushing his head. To make Mary pay for her actions, a local blacksmith decided to try to shoot her dead but was unsuccessful because her hide absorbed the bullets with little effect. Therefore it was decided that she would have to be hanged. A derrick chain was looped around her neck and a railcar-mounted industrial crane lifted her up. As if this wasn't cruel enough, someone had forgotten to release the ankle chain binding her to the track, leaving her suspended for a moment. While hanging there, her ankle tendons tore and then the chain around her neck snapped, sending her crashing to the ground and shattering her hip. Another attempt was successful. A crowd of 2,500 people, including children, gathered to watch.

Maudine Ormsby (Cow)

Maudine, a Holstein with big brown eyes and a sweet disposition, was nominated by Ohio State University's College of Agriculture and surprisingly named 1926 homecoming queen. In the homecoming parade she rode through town on the back of a float with a crown on top of her head; however she was banned from attending the homecoming dance by OSU officials. It seems there had been some electoral fraud—there were 12,000 votes cast in the school with less than 10,000 students. Maudine was actually the runner-up, but took the title because the actual winner, Rosalind Morrison, chose to bow out due to the election fraud. Maudine Ormsby's memory lives on at OSU, and at the student union there is a conference room named in her honor.

Max *see* Faro

Max *see* Kaiser

Max (Dog)

Max was a collie mix who served in the war with three armies. He started as a Russian messenger/sentry dog, then was captured by the Germans and then the British. He understood commands in all three languages. After the war he was adopted by an ex-officer and taken to London.

Michael (Mike) (Dog)

Mike the Airedale was trained by Colonel Richardson to be the first English messenger dog during World War I. Working for the advancing forces in France, Mike ran under constant fire to bring messages to and from the advancing forces to headquarters. Fortunately Mike was never wounded, and when the forces were demobilized, he was brought back to England by Captain R.R. Slater to live out his life to the age of sixteen. One affliction Mike had from the war was crouching or hiding when he heard loud bangs; otherwise he was happy.

Michael (Dog)

Michael falls into the category of "other dogs" whose specific job in the war was not classified. Michael's owner was left for dead in "no-man's land" but was really alive, just unconscious. Michael dragged him back to the trenches and saved his life.

Midnight (Horse)

A coal-black foal with a single white star on her forehead was born on October 31, 1905, and named Midnight. She was raised and trained by Guy Haydon, and a very strong relationship formed between them. When World War I began, Lieutenant Haydon left the front lines on Gallipoli on a ship bound for Egypt to reunite with Midnight. On October 31, 1917, in Beersheba, the last great cavalry charges in history were led by two Australian brigades. At the Battle of Beersheba, Midnight died taking a bullet meant for his rider, Lieutenant Haydon, therefore saving his life.

Midnight (Horse)

Midnight, despite his thoroughbred and Percheron bloodlines, held the world's bucking championship title for fourteen years. This black horse first belonged to Jim McNab, owner of the Cottonwood Ranch near Alberta, Canada. McNab had marked Midnight to be his horse, and after months of winter grazing, Midnight gave his owner some difficulty saddling; however McNab was not prepared to be thrown off Midnight three times. After a few weeks of showing his horse kindness, McNab was allowed to saddle and ride Midnight. After McNab let Twenty-One Johnson (who made his living traveling to ranches collecting bets from cowhands) mount his horse, Midnight refused to let McNab ride him. It seemed as if Midnight felt that McNab had betrayed their friendship by letting a stranger ride him. Verne Elliot, who supplied rodeo shows with equipment, men, and animals, purchased Midnight, who became his star performer—touring England, Canada, and the United States. Midnight astounded audiences with his gentleness after throwing his rider off his back in as little as three jumps. Instead of stampeding, he would nose the rider and carefully step over him on his way out of the arena. Midnight's kindness in spite of being a bucking champion mystified everyone.

Mike (Chicken)

Mike became a headless chicken on September 10, 1945, when his number came up and he was destined for the Olsen family dinner table. But he survived his execution: the axe missed his jugular vein, leaving him with one ear and a semi-attached brain stem intact. Mr. Olsen continued to care for Mike by feeding him a mixture of water and milk with an eyedropper. Mike toured the country, making his owner approximately $4,500 a month (over $70,000 in today's economy), and was featured in *Life* and *Time* magazines. On March 17, 1947, at 15 and a half months old, Mike choked to death, alone in his hotel room in Phoenix, Arizona. Since 1999 on the third weekend of May, an annual event, "Mike the Headless Chicken Day," has been held in his hometown of Fruita, Colorado. The event includes a "5K Run Like a Headless Chicken Race," "Egg Too," "Pin the Head on the Chicken," the "Chicken Cluck-Off," and "Chicken Bingo."

Mila (Beluga Whale)

During a diving contest in July 2009 at Polar Land, an aquarium in the city of Harbin, China, twenty-six-year-old Yang Yun was hoping to become a trainer. Contestants dove down into an icy cold whale pool without any breathing apparatus, and whoever stayed down the farthest and longest won. Yang Yun's turn came and she dove about twenty-five feet when her legs cramped up and she couldn't move and sunk deeper. Two Beluga whales, Mila and Nicola, came swimming to see what was going on. Mila grabbed hold of Yang Yun by one ankle and started pushing her back to the surface. Fortunately Belugas have tiny teeth, so Yang Yun was not injured by her rescuer.

Minnie (Horse)

During operations in Burma in 1944, a pack pony unexpectedly gave birth to a foal named Minnie. The arrival of the foal was a complete surprise because there was no record of a pregnant horse. The foal was born during a Japanese attack, with bombs exploding all around. When the troops heard of the new member, they were all interested in Minnie's progress. During heavy bombing, some mules broke loose and kicked Minnie above the right eye. When medical treatment was given, the eye was saved. As Minnie's condition improved so did the morale of the troops. She was too weak to travel with the troops, but when she was better she could be seen drinking tea from a pot. On November 8, 1951, Minnie died of pneumonia in Ismailia. Before she was interned, her tail and four hooves were removed. Two hooves were made into an inkwell and paperweight and were displayed along with her tail and ceremonial blanket in the Regimental Museum in Bury. Her other two hooves were given in remembrance to the boroughs of Bury and Rochdale.

Miss Baker (Monkey)

Miss Baker, a squirrel monkey born in 1957, was purchased along with 25 other

America's first lady of space, Miss Baker, a squirrel monkey, was one of two animals launched into space to return alive. She went up in a *Jupiter* IRBM in 1959 (NASA/Marshall Space Flight Center).

squirrel monkeys in Miami, Florida. She was one of two animals in the United States that were launched into space and returned alive. NASA called her "America's first lady of space." Miss Baker awaited the flight tucked into a shoebox. On May 28, 1959, at 2:39 a.m., a *Jupiter* rocket lifted Miss Baker and a rhesus monkey, Miss Able, to an altitude of 300 miles and an acceleration of 38 Gs, for 16 minutes with 9 minutes of weightlessness. Post-flight, Able and Baker appeared on the June 15, 1959, issue of *Life* magazine. In 1962, Miss Baker "married" Big George (who died on January 8, 1979). In 1971, Miss Baker moved to the U.S. Space and Rocket Center in Huntsville, Alabama. While there she entertained visitors at the museum and received 100 to 150 letters a day from children. In 1979, the mayor of Huntsville, Alabama, proclaimed her twenty-second birthday, June 29, "Monkeynaut Baker Day." That same year Miss Baker "wed" Norman from the Yerkes National Primate Research Center. On the 25th anniversary of her space flight, Miss Baker received a rubber duck and was treated to strawberry gelatin with bananas. On November 29, 1984, Miss Baker died at age 27 as the oldest living squirrel monkey. She died of kidney failure and is buried on the grounds of the U.S. Space & Rocket Center.

Miss Hap (Kitten)

Marine Sergeant Frank Praytor adopted this two-week-old orphan kitten during the Korean War. He named her Miss Hap because "she was born at the wrong place at the wrong time." It was said that her mother was killed by a mortar barrage, but in truth she was shot by a marine because of her yowling. Miss Hap was left behind in Korea when Sergeant Praytor left, and she became the mascot in the Division PIO office. It is not known whether her second guardian, Corporal Conrad Fisher, was able to take her home,

but Sergeant Praytor said, "I like to think he did."

Mister Ed *see* Bamboo Harvester

Mr. Magoo (Mongoose)

A memorial editorial that ran in the *Duluth News Tribune* on January 10, 1968, sums up the impact Mr. Magoo had in the United States: Titled "Our Mongoose," it read in part, "Death has taken an inhabitant of Duluth very widely known in recent years—in some circles, surely. The mongoose, Mr. Magoo, won international attention. He was the occasion of high-level federal executive-action. President Kennedy saw in the acts which saved the life of the mongoose a classic example of government by the people." In 1962, Director Lloyd Hackl of the Duluth Zoo in Minnesota acquired Mr. Magoo when a foreign ship from India docked and a sailor donated his pet mongoose. A tea-drinking mongoose makes the weasel-like animal seem tame; however in its natural habitat in India it is known for its quick agility in killing cobra snakes. It turns out that due to a 1909 statute of the U.S. Fish and Wildlife Service, it is illegal to keep mongooses in the United States. When federal officials ordered that Mr. Magoo be put down, the people of Duluth mobilized to save his life. Petitions were signed, court orders were sought, and citizens contacted the press and Congress. President Kennedy even became involved in the cause, winning the animal thousands of supporters; Mr. Magoo received a temporary stay of execution. Secretary of Interior Stewart L. Udall had signed a reprieve allowing the mongoose to stay at the zoo as long as he remained popular; otherwise he would be deported to India. On April 19, 1963, the Department of the Interior granted asylum to Mr. Magoo. The mongoose died on January 9, 1968, of old age. Mr. Magoo was mounted by a taxidermist and put on display at the zoo. Basil Norton, zoo director, said of Mr. Magoo, "He had a pleasant disposition right up to the time of his death"; "He had that certain spark that give some animals a special appeal"; and "Another mongoose could never take his place in the hearts and affections of the Duluth people."

Mitz (Marmoset)

Leonard and Virginia Woolf's pet marmoset, Mitz, was mentioned in several sources, including Leonard's autobiography. Quentin Bell (Virginia Woolf's nephew and biographer) compared Mitz to Joseph Goebbels and remembered the creature as being in a perpetual state of vicious fury. Bell wrote, "One always hoped for a severe frost that would finish Mitz." Mitz was babied by Leonard, with whom the jealous monkey was madly in love. During a drive through Germany in May 1935, Mitz kept the Woolfs from harm; they were able to safely pass through the country, which was especially remarkable since Leonard was Jewish. Arriving in Bonn in their open car with Mitz perched on Leonard's shoulder, they found the main road lined with Nazis awaiting the arrival of a dignitary. At the sight of Mitz the crowd shrieked with delight. Throughout Germany people were so excited to see "*das liebe kleine ding* (the dear little thing)." Leonard recalled in his autobiography, "It was obvious to the most anti-semitic storm troopers that no one who had on his shoulder such a 'dear little thing' could be a Jew."

Mocker (Pigeon)

Mocker was another successful pigeon who flew with one eye destroyed. According to reports, Mocker had recovered from his injury.

Modestine (Donkey)

Robert Louis Stevenson bought Modestine for 65 francs and a glass of brandy in the French town of Le Monastier. The donkey was to carry his gear on the casual journey south that he describes in his book *Travels with a Donkey in the Cevennes* (1879). In the book, he describes humorous descriptions of his stubborn and manipulative travel companion, Modestine. For twelve days, they struggled up and down mountains and got lost for over 120 miles. Their slow pace frustrated and angered Stevenson, but as the journey progressed he grew attached to Modestine. At the end of the trip Modestine was worn out and Stevenson sold her.

A quote from the book:

> I had lost Modestine. Up to that moment I thought I hated her; but now that she was gone…. For twelve days we had been fast companions; her faults were those of her race and sex; her virtues were her own.

Moko (Dolphin)

Moko, short for Mokotahi, was a male bottlenose dolphin who spent time with humans on the east coast of the North Island of New Zealand from 2007 to 2010. In March 2008, Moko was seen helping two pygmy sperm whales swimming toward the shore and probably to their death. Moko swam out to the whales and somehow communicated with them to follow him along the beach and back out to sea. During the winter of 2009, a New Zealand swimmer started playing with Moko. When the woman tired, he kept playing and did not allow her to return to the shore. People on the shore heard her panic and went out to rescue her. Moko was just playing and probably lonely; the woman admitted it was too late in the season and Moko meant no harm. Scientists worried about Moko after finding scarring by boats and fish hooks. Moko was found dead on

July 7, 2010, on a beach at Matakana Island near Tauranga.

Monsieur Grat (Dog)

After he resumed living in a small town in Holland in 1644, René Descartes would take his dog, Monsieur Grat (AKA Mister Scratch), on his walks. Some find it remarkable that he had a dog for company because of his view of animals. He asserted, "The greatest of all the prejudices we have retained from our infancy is that of believing that the beasts think." In his view, they cannot reason, therefore they have no souls. Their cries are not emotional but the workings of an intricate machine—"The animals act naturally and by springs, like a watch."

Montauciel (Sheep)

Montauciel means "ascend into the sky" in French. As early as the 1700s, after launching several empty hot air balloon flights, people became curious if humans could survive time spent at high altitudes. In 1783, Montauciel the sheep, along with a duck (a bird used to flying at high altitudes) and a rooster (a flightless bird), took off on the first hot air balloon flight containing a land animal. The Montgolfier brothers launched the balloon at the royal palace in Versailles before King Louis XVI of France and Queen Marie Antoinette. All the animals returned unharmed; therefore scientists concluded that if animals could survive the flight then so could humans. Montauciel was adopted by Marie Antoinette and spent his life eating candy and marshmallows.

Moustache, Mous (Dog)

Moustache was a black French poodle born in September 1799 at Falaise in Normandy, France, and is said to have taken part in both the French Revolutionary and Napoleonic wars. His story, reported in many publications, may be partly fictionalized. Two of the earliest stories were written by Arna Cano in *The Kaleidoscope* magazine of Liverpool in January 1826, and also an account by Alain de Fivas, which was published in 1864. There seem to be two occasions where Moustache proved himself to be a hero. On one occasion, he was following a grenadier regiment through the Italian Campaign of the Revolutionary Wars, and his barking alerted the regiment of a surprise night attack by Austrian forces at their camp near Belbo. There are other sources that place this event in Alexandria, Egypt, during the Egyptian campaign against the Ottoman Empire. A second time, in 1805 at the Battle of Austerlitz, Moustache discovered the presence of an Austrian spy and sustained severe wounds from an artillery blast, from which he had a leg amputated. During the Battle of Marengo, he lost an ear saving the life of an officer by battling with a German corporal's pointer that was attacking the officer. Moustache's life took a turn in 1809 at the Battle of Aspern-Essling, where he met a female poodle. They were happy together for a year and Moustache even became a father. One day a chasseur reportedly gave "him a blow with the flat side of his sabre"; this cruelty proved too much for Moustache and he abandoned his family and regiment. Afterwards he found himself following some dragoons to Spain. There he was useful by giving notice of anything seeming suspicious. Again he proved his heroism by safely bringing to camp the horse of a dead dragoon from the battle of the Sierra Morena. Shortly after, he was stolen and held in captivity by a colonel who wanted to own him. Moustache managed to escape through an open window. After his escape, he fought in some battles at Badajoz during the Peninsular Wars. During these battles, on March 11, 1811, he was killed by a cannonball. The inscription on his grave reads, "Here lies the brave Moustache." Moustache received a medal for his bravery and dedication.

Mrs. Chippy (Cat)

Mrs. Chippy, a tiger-striped tabby, joined Ernest Shackleton on board his ship the *Endurance*, leaving on August 1, 1914, from the East India Docks in London for their voyage to the Antarctic. The ship's carpenter and master shipwright, Henry "Harry" McNish, found the cat sleeping in one of his toolboxes and thought it would make a good companion. Shackleton was glad to have a cat aboard to keep rodents under control and to protect provisions. The cat acquired the name Mrs. Chippy by following McNish around the ship like an "over-possessive" wife, and the name stuck despite discovering that "she" was a tomcat. ("Chippy" is a colloquial and affectionate name in Britain for a carpenter.) A stowaway on the *Endurance*, a 19-year-old British man named Perce Blackborow, joined the crew and became Mrs. Chippy's second-best friend. Mrs. Chippy was not fond of the 70 howling Canadian sled dogs that were also on board living under poor conditions. In mid–January 1915, an ice pack trapped the *Endurance*, and a month later the ship's predicament got worse as the Antarctic winter set in. By October 1915, winter seemed to be winding down and the crew had high hopes that the ship would break from the ice and continue its voyage; however the pressure of the ice had damaged the ship. Shackleton made a plan that the crew would escape with three lifeboats and sail towards land 350 miles away. Shackleton's plan only allowed enough room on the lifeboats for the crew and the essentials for survival, which meant no room for Mrs. Chippy or the dogs. Thus the decision was made to shoot the animals. Three and a half months later the crew made it safely to land. In 1958, in honor of the carpenter,

the British Antarctic Survey named a small island off South Georgia McNish Island. In 1959, the New Zealand Antarctic Society put a headstone on McNish's unmarked grave. In June 2004, sculptor Chris Elliott created a life-sized bronze sculpture of Mrs. Chippy atop McNish's grave in recognition of the cat's part on the expedition. In 2006, the library in his birthplace of Port Glasgow unveiled a plaque dedicated to Harry "Chippy" McNish.

Muck *see* Tank and Muck

Muezza (Cat)

The Islamic prophet Muhammad was very fond of cats, particularly one named Muezza. Muhammad was so attached to him that he would let the cat sit on his lap while he gave sermons. There are several tales about Muezza. One says that the cat saved Muhammad from a deadly venomous snake bite. Another story has Muhammad cutting off the sleeve of his prayer robe where Muezza was asleep, rather than waking him. These are all charming folktales about Muezza. In Islam the domestic cat is a revered animal. Felines owe their sacred status to Muhammad and his love of animals.

Muhamed (Horse)

Muhamed was one of three horses raised by Karl Krall in the town of Elberfeld in the 19th and early 20th centuries. Krall had three horses that he claimed had special abilities such as reading and doing arithmetic. The other two horses were Clever Hans and Zarif. The horses were clever (which spurred the term "The Clever Hans" effect). While all of Krall's horses had the apparent ability to read and do basic arithmetic, Muhamed was the most gifted. Reportedly Muhamed could mentally extract the cube root of numbers and also perform music, distinguishing between harmony and discord. Psychologists and scientists tested Muhamed by writing a number on the blackboard and asking Muhamed to extract the cube root. Muhamed used his hooves to tap out the answers; his left foot represented tens and his right foot represented ones. For example, to answer "sixty-five" he tapped six times with his left foot and five times with his right foot. Two scholars tested Krall's horses: psychologist Edward Claparede felt their abilities were genuine, and Belgian writer Maurice Maeterlinck claimed Krall had "humanized" the horses. Muhamed disappeared in World War I while serving as a draft animal.

Munito (Dog)

Munito the "Wonderful Dog," a famous performing poodle, was born in about 1815 near Milan, Italy, and owned and trained by Signor Castelli d'Orino. Munito's father was a hound and his mother was a water-spaniel. Beginning his education at 2 and a half months and lasting thirteen months, Munito then toured Europe from the 1810s to the 1830s. During training he never left the house, nor did he see anyone other than his trainer.

His daily exhibitions in Mr. Saxton's room in London amazed everyone who saw him perform his many talents:

This matchless dog, who understands equally well the French and Italian languages, shows every evening new talents before the numerous and select company which his reputation brings to him: he not only proves that he knows all the cards of a complete pack, all the letters of the alphabet, all the numbers of arithmetic, all the points of the dominoes, and all the shades of colors, by bringing them at command; but, besides, he guesses and brings the cards which being drawn from a pack by different persons, and shuffled by others, are afterwards spread covered on the carpet. He looks at the color of a dress, and brings to his master a piece of the same shade, if

there is any among the samples before him; if not, the nearest to it. He reads all sorts of hands; but not being endowed with the gift of speech, he, with printed letters placed at his disposal, makes copy of the words presented to him; afterwards, he plays a game at dominoes with whoever will condescend to procure him that little pastime.

An account of the training follows:

That after four months study, he knew cards, letters, figures, colors, and brought them when commanded; that the nine last months have been employed in making him combine letters, figures, and dominoes; and, finally, that in all the course of his lessons, his master has continually employed with him the rules which the wisest and most learned writers have given on the education of children; that he has never struck him, nor spoke to him angrily; that he has always adopted towards him a progressive course, founded upon reason, proportioned to his means, and supported by caresses and rewards, to make him do what he wished; so that, with good method, some chestnuts, and gentle treatment, he taught him, and makes him repeat every day, what he knows.

Both Charles Dickens and Jules Verne claimed that they discovered some of the trainer's secrets through treats, certain smells or sounds.

Murphy (Donkey)

Shortly after landing on Gallipoli in 1915, Private John Simpson, who served in the 3rd Australian Field Ambulance, befriended a desert donkey, Murphy (AKA Abdul). Carrying wounded soldiers on his back, Murphy, accompanied by Simpson, transported them to the dressing station. Traveling unarmed amidst shrapnel and gunfire across front lines day and night, the pair transported dozens of Australian soldiers to treatment. Murphy was awarded the Purple Cross for heroism. According to legend, Simpson and Murphy saved 300 lives at Gallipoli.

Mutt (Dog)

The best-loved memoir book *The Dog Who Wouldn't Be* (1957) by Farley Mowat tells the story of a black and white mutt with original ways. He was bought as a puppy in 1929, when 8-year-old Farley lived in Saskatoon, Saskatchewan, with his mother and father, a librarian. Mutt learned to go after birds by diving, running them down, or climbing a tree. He climbed trees in town, too, as well as ladders to chase cats. One day he retrieved a stuffed and mounted ruffed grouse from a store display to win a bet. He rode passenger in an open car wearing goggles. Mutt couldn't seem to leave a skunk alone.

Myobu No Omoto (Cat)

Omoto, the "Lady-in-Waiting," was so loved by Ichijo, emperor of Japan (986–1011), that when she had to be rescued from a dog, the emperor exiled the dog and sent the owner to prison. When domestic cats were introduced in China, only the emperor and noblemen could afford to keep them. The cats were called *Kara neno*, which means "Chinese cats." When a mother cat gave birth to a litter of five kittens at the palace, the day was so important that the date, nineteenth day of the ninth month of the year 999, was recorded. The emperor ordered special royal attendants to feed the kittens, and he even dressed the animals in clothes. He gave orders that a wet nurse take care of the kittens as if they were his own children.

Mysouff (Cat)

This cat lived with Alexandre Dumas, author of *The Three Musketeers*, and Dumas' mother. Mysouff ought to have been a dog. When Dumas went to work in the morning, Mysouff accompanied him to a certain corner and was there to greet him in the evening. He seemed to be clairvoyant by knowing Dumas' timing. Mysouff

was succeeded by Mysouff II years later. By that time, Dumas also had a dog, three monkeys, and birds. One day the monkeys let the birds out of their cage and Mysouff ate them all. Mysouff was sentenced to five years with the monkeys in their cage. When Dumas needed money, he sold the monkeys and Mysouff regained his freedom.

Mzee *see* Owen and Mzee

Nancy (Springbok)

Nancy was the famous mascot of the 4th South African Infantry Regiment. In March 1915, Nancy (a springbok, or southern African gazelle), was just over a year old when her owner, Mrs. McLaren Kennedy of the farm Vierfontein in the Orange Free State, volunteered her pet for war service. Bugler Private A.E. Petersen was put in charge of Nancy's training, and for the next six months she was taught to respond to regimental calls and to be on her best conduct in parades and ceremonies. Her unit was stationed at Mex Camp in Alexandria, and one morning Nancy went missing and by midday she was considered AWOL. The following morning she was still AWOL and a house-to-house search began. On the third day, a group of bagpipers marched into the desert in different directions, and fortunately Nancy recognized the sound of the pipes and returned to camp. During the Battle of the Somme, a shell exploded near Nancy, and she broke her left horn on a wall as she fled. From this point on, her horn started to grow downward at an angle. On November 26, 1918, Nancy died after catching pneumonia. Nancy was buried with full military honors in the cemetery of the village of Hermeton-sur-Meuse. Her skin was stuffed and is displayed at the South African National Museum of Military History in Johannesburg.

Nansen

122

Nansen (Cat)

Nansen, a black and white kitten, was brought on board the Belgian ship RV *Belgica* by Norwegian cabin boy Johan Koren. The ship set sail for Antarctica in 1897 on a scientific expedition under the command of Captain Adrien de Gerlache. The cat was named after Norway's Antarctic explorer Fridtjof Nansen. The crew became very fond of the cat; however, becoming trapped in the ice, the ship was forced to wait out the cold and darkness of the winter. Nansen's personality drastically changed to the point where he seemed almost deranged. He was ill at ease and became angry and unfriendly if disturbed. Nansen's cause of death on June 22 is unclear. Nansen is featured in a book, *Through the First Antarctic Night*, written by the ship's doctor, Frederick Cook.

Ned *see* Tusco

Neddy *see* Jimmy

Nell (Dog)

Nell was a high-strung show collie and not the type of dog one would associate with being a messenger dog during World War I. However, she turned out to be a reliable messenger dog that saved hundreds. After the war, she happily retired with her handler.

Nellie (Dog)

Nellie falls under the category of "other dogs" whose specific job in the war was not classified. Nellie was a fox terrier who followed her owner into the first battle of Ypres. Later, she was adopted by a Belgian regiment and wounded twice by shrapnel. A Belgian missionary brought her over to America, where she lived out her life.

Nellie Jay *see* Elm Farm Ollie

Nelson (Cat)

Winston Churchill had a reputation of being a great animal lover. Nelson, a big grey tom, was Churchill's best-known cat during the war years. Churchill recalled the time when Nelson demonstrated his true bravery when he saw him "chase a huge dog out of the Admiralty" (the Executive Department formerly having general authority over British naval affairs). That was when Churchill decided to adopt the cat and name him after Admiral Horatio Nelson. Nelson even went to live at Churchill's residence, Number 10 Downing Street in London, where he bullied and chased the resident cat "Munich Mouser" out of Number 10 completely.

Nemo A534 (Dog)

Nemo, a German Shepherd born in October 1962, served in the United States Air Force during the Vietnam War with his handler, Airman Bryant. In January 1966, the two were transferred to the Republic of South Vietnam, where Bryant was stationed at the 377th Security Police Squadron at the Tan Son Nhut Air Base. Six months later, 22-year-old Airman 2nd Class Robert Thorneburg became Nemo's new handler. On December 4, 1966, the air base was attacked by the Vietnamese. During the attack Nemo suffered a bullet that entered his right eye and exited through his mouth. Thorneburg was also wounded; Nemo saved his handler's life by crawling on top of his body and guarding him until medical help came. On June 23, 1967, Nemo returned to the United States with the honor of being the first sentry dog officially retired from active service. Upon his return to the United States, he was given a retirement kennel at the Department of Defence Dog Center in Texas, but he continued to work as a recruiting dog until his death in December 1972. Of the more than 4,000 dogs serving in Vietnam, only about 200 returned home.

Nils Olav (Penguin)

Nils Olav was a king penguin, the second-largest kind of penguin, that resided at the Edinburgh Zoo in Scotland. This esteemed penguin was personally knighted by the King of Norway in 2008, and was the mascot colonel-in-chief of the Norwegian King's Guard. Each time the guard visited Nils Olav in Edinburgh he was promoted, and when he outranked almost everyone he began to be referred to merely as "mascot" in Norway. An article on the Listverse website reasoned, "After all, no one wants a situation where they have to figure out how to obey a command that goes 'Waark!'"

Nim Chimpsky (Chimpanzee)

Nim (full name Neam Chimpsky), a chimpanzee born on November 19, 1973, was a subject (codenamed 6.001) of a study of animal language-acquisition. The study was at Columbia University and led by Herbert S. Terrace with psycholinguist Thomas Bever, who led the linguistic analysis. Project Nim was an investigation similar to an earlier study by R. Allen and Beatrix Gardner called Project Washoe, which involved a chimpanzee being raised like a human child. Project Nim was essentially a study to challenge Noam Chompsky, a leading theorist on human language structure and generative grammar, on his thesis that "only humans have language." Both chimps, Nim and Washoe, were able to use American Sign Language to communicate. The study's focus was on "Nim's ability to make different responses to different sequences of signs and to emit different sequences in order to communicate different meanings." Terrace concluded that "the chimpanzee did not show any meaningful sequential behavior that rivalled human grammar." After the experiment ended, Nim went back to the Institute for Primate Studies in Oklahoma. Later, Nim was sold to the Laboratory for Experimental Medicine and Surgery in Primates. Nim was rescued by the Black Beauty Ranch, where he lived until his death on March 10, 2000. *Project Nim*, a documentary by James Marsh (produced by BBC Films, Red Box Films and Passion Films), opened the 2011 Sundance Film Festival.

Nipper (Dog)

Nipper was so named because he would nip at the backs of visitors' legs. This black and white mixed-breed terrier, probably part Jack Russell terrier, born near Bristol, England in 1884, is the RCA dog. He was adopted by artist Francis Barrand after Francis' brother Philip, the dog's first owner, died. In the 1890s, Barrand was inspired by the way Nipper cocked his head at the sounds coming out of the gramophone. Barrand painted the scene in *His Master's Voice* three years after Nipper's death in 1895 from natural causes. After Barrand replaced the gramophone with a more recent turntable, he was able to sell the painting to numerous audio brands. Barrand kept selling paintings. In 1901, the Victor Talking Machine Company of New Jersey (later merging with RCA) acquired the American rights to the painting and adopted it as its trademark. In commemoration, on March 10, 2010, a small road near his final resting place in Kingston was named Nipper Alley. A four-ton statue of Nipper appears on the roof of the RCA building in Albany, New York. A second, smaller statue is on top of the Maryland Historical Society building in Baltimore. Another small statue of Nipper sits above a doorway in the Merchant Ventures Building in Bristol. The music video of Cyndi Lauper's song "Time After Time" features a life-size ornament of Nipper.

Nipper (Ferret)

Known for their love of burrowing, ferrets can put their skills at running through

Francis Barrand's 1898 painting of Nipper, the RCA dog, titled *His Master's Voice*.

pipes to a variety of professional uses. When wires cannot be pushed through tubes or tunnels with rods, ferrets are known for pulling wire through underground tunnels to help lay the wire. In 1977, camera cables had to be led by a draw wire through a six-inch underground duct to equipment for the Queen's Silver Jubilee. The draw wire was missing, Nipper was called upon to replace the wire. Equipped with the wire at one end of the duct, Nipper scurried to his reward of bacon at the other end, 75 yards away. The nylon thread attached to his harness was used to pull through the duct the first of increasingly larger lines. Four years later Nipper was harnessed to demonstrate the procedure. He was prepared to perform the procedure, if needed, at the royal wedding of the Prince of Wales and Lady Diana Spencer in 1981.

No. 2709 (Pigeon)

Number 2709 was a British bird. On October 3, 1917, she was sent with a message to her loft nine miles away. During her flight she was wounded and spent a night in the rain. Her trainer gave her up for dead. The next day she returned to her loft; she had been shot and was missing a leg. Also the cylinder holding the message was driven into her body, which ultimately caused her death.

O-Six (Wolf)

O-Six, a charismatic, unusually large alpha female wolf, was named for the year of her birth. Powerful with gray fur and faint black ovals around her eyes, she was a kind and merciful leader. She first came to the attention of wolf watchers in 2009 as a lone wolf. She wandered for quite some time when she was spotted in Lamar Valley in majestic Yellowstone National Park. She was beloved by wolf watchers and naturalists and became a social media star. She raised her pups and protected her pack even while constantly challenged by hunters, cattle ranchers, and other wolves in

Yellowstone. In 2012, O-Six was killed by a trophy hunter outside the sanctuary of Yellowstone. Her death caused an international outcry comparable to the killing of Cecil the Lion in 2015. She was described as a once-in-a-generation hunter.

Obaysch (Hippopotamus)

Obaysch's name comes from an island on the White Nile where he was captured at less than one year old. He was the first hippopotamus in Great Britain since prehistoric times and the first in Europe since ancient Rome. Obaysch came to the London Zoo on May 25, 1850, after the Ottoman viceroy of Egypt, Abbas Pasha, agreed to swap Obaysch with British Consul General Sir Charles Murray (later known as "Hippopotamus Murray") for some greyhounds and deerhounds. Obaysch became an instant sensation, attracting 10,000 visitors each day, creating a frenzied trade in hippo memorabilia and inspiring "The Hippopotamus Polka" by L. St. Mars. Abbas Pasha also sent a female hippo named Adhela, who arrived in London on July 22, 1854. In 1871, after 16 years, Obaysch and Adhela produced offspring, but the calf died two days after being born. The next year a second calf died, but a third calf, a female born on November 5, 1872, survived and was named Guy Fawkes. Obaysch died on March 11, 1878; Adhela died on December 16, 1882; and Guy Fawkes died on March 20, 1908.

Old Abe (Bald Eagle)

Old Abe (May 27, 1861–March 26, 1881; named in honor of Abraham Lincoln) was a majestic bald eagle and the most famous Civil War mascot. Chief Sky of the Chippewa tribe in northern Wisconsin had spotted a treetop nest of two eagles on a hunting and fishing expedition. To capture them he cut down the tree. One eaglet died and the other became his pet. The McCann

Old Abe, pictured here in 1876, was named in honor of Abraham Lincoln and the most famous Civil War mascot (Liljenquist Family Collection of Civil War Photographs, Prints and Photographs Division, Library of Congress).

family bought Old Abe from the Chippewa for a pet when he was just a young bird, in exchange for a bushel of corn. In August 1861, a company of volunteers from Eau Claire called the "Eau Claire Badgers" bought Old Abe from the McCanns for $2.50. Company C, 8th Wisconsin Volunteer Infantry Regiment, swore him in as their mascot, and a young soldier, James McGinnis, volunteered to care for the eagle. Before the regiment went into battle, Old Abe participated in recruitment events, marches and parades. This heroic eagle took part in 42 battles by flying above and screeching at the enemy but never flying over the lines of either army. Thinking how demoralizing the loss of

Old Abe would be to the enemy regiment, the Confederates tried to capture or kill the eagle. Obviously the plot failed. On June 19, 1864, 240 reenlisted veterans and their eagle went on furlough, arriving in Chicago to waving flags along Madison Street and ringing of the bells of the city. A national salute was given and the veterans were also addressed by several dignitaries. On June 26, 56 veterans of Company C and Old Abe arrived at Eau Claire and were greeted by booming cannons, martial music, patriotic songs and a great feast. In August 1864 the soldiers and Old Abe returned from furlough to northern Mississippi after a large victory by Confederate Major General Nathan Bedford Forrest at Brice's Cross Roads. A month later Officer A.J. Mower's four brigades defeated Forrest near Tupelo, Mississippi, causing numerous casualties. Old Abe's last battle occurred on August 13, 1864, on yet another defeat of Forrest at Hurricane Creek, Mississippi. After the war, Old Abe made many public appearances and proved to be a great fundraiser; for example, to raise money for soldier relief thousands of photographs of the bald eagle were sold. On September 28, 1864, he "retired" and was given to the state of Wisconsin. State officials declared him a "War Relic" and created an "Eagle Department," which included a two-room "apartment." On March 26, 1881, Old Abe died of smoke inhalation when the state capital caught on fire. He was immediately stuffed and put on display. In 1904, in yet a second fire, his stuffed body was destroyed. In his memory a stone sculpture is on display at the top of the Camp Randall Arch.

Old Bet (Elephant)

On April 13, 1796, Old Bet arrived in the Americas on the private armed ship *America*. During the early 1800s, the elephant was put on display all across the Eastern Seaboard. Interestingly Old Bet

was purchased by farmer Hachaliah Bailey (whose ancestral namesake would later be tacked onto the end of the renowned Ringling Bros. and Barnum & Bailey Circus) for pulling a plow around the farm. Bailey realized that he could make more money exhibiting Old Bet as the main attraction of a traveling menagerie and charging a coin or a two-gallon jug of rum for admission to the show. On July 24, 1816, Old Bet was killed by a farmer while on tour near Alfred, Maine. The farmer reasoned that the impoverished should not waste their money on such frivolous things as circuses.

Old Douglas (Camel)

Old Douglas, a one-hump dromedary camel, was a remnant from the U.S. Camel Corps, an experimental program by the U.S Army to develop alternative transportation in the southwestern part of the country. Jefferson Davis, U.S. secretary of war, was a strong advocate of the proposal to experiment with camels. Old Douglas was used by Company B of the 43rd Mississippi Infantry, part of the Confederate Army during the American Civil War. Old Douglas was described as a quiet, peaceful fellow and a general favorite within the regiment. He was intentionally killed by Yankee sharpshooters as he was grazing peacefully on a hillside in the final days of the siege of Vicksburg. His death saved the men of his regiment from starving because they ate him. A headstone commemorating Old Douglas was erected in Vicksburg around 2006. It is believed that his regiment shot the sharpshooter who killed Old Douglas.

Old Drum (Dog)

This black and tan hound's death became the basis of a lawsuit in 1869. Old Drum belonged to a Missouri farmer, Charles Burden. An irate neighbor,

Leonidas Hornsby, believed he had lost sheep to dogs and had his nephew shoot Old Drum. Burden sued Hornsby and won, and then Hornsby appealed and won a reversal. Burden was granted a new trial based on new evidence. Burden's lawyer, George Graham Vest, had a closing argument that centered on the bond between men and their dogs in history, literature, and legend. As a result Burden won the suit and was awarded $50 (equivalent to over $1,000 today), and Vest's one-hour speech has become a classic. A memorial monument stands near where Old Drum's body was found. In 1953, a bronze statue of Old Drum with the text of Vest's "Eulogy on a Dog" written on the base stands at the Warrensburg, Missouri Courthouse Square, and a bust of Old Drum is in the Missouri Supreme Court building in Jefferson City. In 2017, the Missouri State Senate officially designated Old Drum as "Missouri's Historical Dog." Vest's speech has been depicted in movies set in Missouri involving sheep owners who shoot the dog but are in turn shot by the dog's owner, who is then tried for murder.

Old Harvey (Dog)

Old Harvey, a white bulldog, served as a mascot for the 104th Ohio Volunteer Infantry. A Newfoundland dog, a cat, and a tamed raccoon also served as mascots for this unit.

Old Henry Clay (Horse)

Old Henry Clay, also called "Americas' National Thoroughbred Trotting Horse" or "Father of American Trotting Horses," was foaled in 1837 and purchased by Colonel William W. Wadsworth of Geneseo in Livingston County, New York. After the horse's trotting race days were over, he was used for breeding until his death in 1867. On April 22, 1881, fourteen years after his death, his skeleton was exhumed and donated to the United States National Museum. All that remains of Old Henry Clay is his mandible and part of his skull; they are kept in the research collection at the Smithsonian Museum Support Center in Suitland, Maryland. Sired from Andrew Jackson, a noted trotting horse, Henry Clay became the foundation sire of the Clay strain. Existing during the 1900s, Clay trotting horses no longer exist.

Old Ike (Sheep)

Old Ike, a tobacco-chewing ornery Shropshire ram, and thirteen ewes were pastured on the White House grounds during World War I to trim the lawn and save manpower. In addition to trimming the lawn, they worked on shrubbery and beds of perennials. The sheep were sheared regularly and the wool was auctioned off by Woodrow Wilson. It was reported that $52,823 of the proceeds were donated to the Red Cross in 1919. In 1920, the flock was given to Mr. Probert's farm in Maryland. Old Ike was given his last chew in 1927. Wilson was the first president to use farm animals to help the country's finances.

Old Jack see **Jack Brutus**

Old Reliable see **Gordo**

Old Shep (Dog)

Fort Benton, Montana, is not on the map where it should be located, but upon exiting the highway, a traveler will find a kind of oasis with full-grown trees and a town with well-kept buildings. It is the home of the "Smithsonian Buffalo," which also includes Hornaday's Bull, the feisty bison Sandy, and Montana's official Lewis and Clark Memorial. Fort Benton's most famous resident is the "forever faithful" sheepdog, Old Shep, whose legacy began in the summer of 1936. When Shep's owner,

Old Ike, a Shropshire ram, with 13 ewes on the White House grounds in 1919 (Library of Congress).

a sheep herder, went to Fort Benton seeking medical treatment after falling ill, Shep went along. A few days later, the sheep herder died, and Shep watched his master's body as it was loaded onto the train heading back east to his relatives. For the next five and a half years Shep went to the train station to greet the four trains that arrived each day, waiting for his master's return. When Old Shep was featured in *Ripley's Believe It or Not*, he became quite a sensation, receiving fan mail and Christmas gifts. Train travelers took detours to Fort Benton just to see Shep meet their train. On January 12, 1942, being arthritic and deaf, Shep did not hear the 10:17 a.m. train and slipped on an icy rail and died. Shep's funeral was attended by hundreds; he was buried in a plot looking down at the station, and a painted wooden cutout was placed next to his grave. His grave became neglected when the train quit coming through Fort Benton. New fans refurbished his grave and monument, and in 1994, a larger-than-life statue was unveiled by the town and Shep's gravesite became a detour attraction.

"Forever Faithful–Old Shep," field review by the team at RoadsideAmerica. com, *Your Online Guide to Offbeat Tourist Attractions*

Also see: "The Noble Dog Mausoleum" at RoadsideAmerica.com

Old Sorrel *see* **Little Sorrel**

Old Whiskers (Goat)

President Benjamin Harrison's son Russell had an ornery pet goat named Old Whiskers (AKA His Whiskers). Benjamin loved his grandchildren and allowed Old Whiskers to stay at the White House so they could play with the goat. Old Whiskers used to pull a cart in which the grandchildren would ride. One day the ornery billy goat ran through the White House pulling the grandchildren behind. President Harrison, decked out in a top hat, was seen waving a cane and yelling at the goat while chasing him in hot pursuit.

Old Whiskers pulls President Benjamin Harrison's grandchildren in 1891. From left: Marthena Harrison, Harrison's son Russell Benjamin Harrison, Benjamin "Baby" McKee, and Mary McKee. The dog is Dash (Frances Benjamin Johnston Collection (Library of Congress).

Old Whitey (Horse)

Old Whitey was the favorite mount of General Zachary Taylor, who was called "Old Rough and Ready." General Taylor, in his farmer's straw hat, and Old Whitey were a familiar sight during the Mexican-American War. When Taylor became president in 1849, his old war horse grazed on the White House lawn. Before the Secret Service existed, visitors gave him sugar and carrots or pulled hairs out of his tail for souvenirs. When President Taylor died in 1850, Old Whitey marched in the funeral procession. Historian Barry Landau, author of *The President's Inauguration: An American Pageant*, said of Old Whitey, "When he would hear parade music, he would start prancing and would want to get into the action." Not much is known about what happened to Old Whitey except that he died of old age.

Old Whitey (Horse)

It is said that Old Whitey and Major Russell Hastings of the Union Army together survived 19 battles. Old Whitey was also owned by President Rutherford B. Hayes and was the preferred horse of "Mother" Bickerdyke, a well-known battlefield nurse.

Olga (Horse)

Olga, a bay mare, was one of 186 horses that were part of the Metropolitan Police Mounted Branch in London during World War II. These horses were used to aid civilians during the Blitz and through-out much of the war (from September 1940 to late 1944). Olga and her handler, P.C. J.E. Thwaites, served in crowd control and rescue in south London near Tooting. They were patrolling Besley Street when a bomb exploded 300 feet in front of them, destroying four houses, killing four people and causing a plate-glass window to fall in front of Olga. After settling down, she helped divert sightseers, allowing Thwaites to administer help. On April 11, 1947, Olga was awarded the Dickin Medal at a ceremony at Hyde Park.

Oliver (Chimpanzee)

After Oliver's death on June 2, 2012, his handler, Ralph Helfer, said, "He was just on a different level. He had very human-like traits. He loved coconut sorbet and that food got the biggest hoots and hollers. But if he didn't like something, he'd hand the bowl back to you, like the time he tried sugar-free pistachio pudding." Oliver was a common chimpanzee whose human-like characteristics led to an interesting life story. He had a flat face and was bipedal—unlike other apes that walk on their knuckles. Trainers Frank and Janet Berger acquired Oliver in the 1960s and raised him in their home until he was 16, when Berger claimed that they had to sell him because he had become sexually attracted to her. Oliver was sold to Ralph Helfer, owner of Enchanted Village in Buena Park, California, where he became an exhibit featured as a human-ape hybrid or a new subspecies. After living at a scientific and cosmetic facility, in 1998 he was transferred to Primarily Primates in Bexar County, Texas, where he lived in a spacious and open-air cage. A geneticist from the University of Chicago tested his DNA and found that Oliver had a normal chromosome count for chimpanzees. His other unusual traits—cranial morphology, ear shape, freckles, and baldness—fell within the range of variability in the common chimpanzee. After his death his body was cremated and his ashes were scattered over the Primarily Primates grounds. In the 2011 film *Rise of the Planet of the Apes* the character of Caesar was based on Oliver.

OR-7 (Wolf)

OR-7 (AKA Journey), is a grey male wolf born in April 2009 in Oregon state, the first confirmed wild wolf spotted in western Oregon since 1947 (none had been spotted in California since 1924). In February 2011, the Oregon Department of Fish and Wildlife (ODFW) attached radio collars on several wolves to study their migration. The wolves were assigned numbers; OR-7 represents the seventh wolf to be collared. In September 2011, most likely in search of a mate, OR-7 left the Imnaha pack in the Wallowa Mountains. By late December he had made his way into northern California. From there, he returned to Oregon in March 2013; he had traveled over 1,000 miles. OR-7's migration caught the world's attention, and in 2012 the nonprofit group Oregon Wild sponsored a children's contest to name OR-7—the winning name was Journey. In May 2014 cameras spotted OR-7 with a possible mate. A month later, the United States Fish and Wildlife Service and the ODFW took fecal samples of two pups for DNA testing. The DNA confirmed that the pups were the offspring of OR-7 and his mate, which increased the probability for a long-term wolf population in California. Therefore in June 2014, in a 3–1 vote by the California Fish and Game Commission, the wolves became protected under the state Endangered Species Act. The batteries in OR-7's collar eventually expired, so camera and live sightings became the only way to track OR-7. By 2017, the Oregon wolf population reached 112. German filmmaker Clemens Schenk made a documentary, *OR-7: The*

Journey, starring a look-alike wolf from an Idaho reserve.

Orangey (Cat)

Orangey (AKA Jimmy or Rhubarb), a talented orange tabby cat, was by far the most famous feline star of the silver screen in the 1950s and early 1960s. He was owned and trained by Frank Inn, a Hollywood animal trainer. Orangey was the only cat to win two Patsy Awards (the animal equivalent to the Oscar). The first Patsy was for his role in *Rhubarb* (1951), about a cat who inherits a fortune. His second Patsy was for his role of "Cat" in the movie *Breakfast at Tiffany's* (1961). He has also been credited as the cat in the film *The Diary of Anne Frank* (1959), and as the family pet in the film *The Incredible Shrinking Man* (1957). Orangey also had a regular role as "Minerva" on the television series *Our Miss Brooks* (1952–1958). Reportedly Orangey's personality was the opposite of his charming on-screen characters. One studio executive was quoted as saying that Orangey was "the world's meanest cat." He was also mischievous, clawing the actors and leaving the set and hiding out.

Oscar (Cat)

"The Feline Grim Reaper," Oscar was raised in the dementia unit at Steere Nursing and Rehabilitation Center in Providence, Rhode Island. He became famous from an article written by Professor David Dosa of Brown University about his extraordinary ability to predict death. Oscar would visit the beds of the terminally ill patients, napping with them, and within hours they would pass away. When he was found in someone's bed, the doctors and nurses would urge the family to come say goodbye to the patient. By 2010 he had predicted more than 50 deaths.

Oscar or Oskar *see* Unsinkable Sam

Owen (Hippopotamus) and Mzee (Aldabra giant tortoise)

Owen and Mzee became the subjects of attention after forming an unusual bond of friendship at Haller Park in Kenya. Following a December 2004 tsunami, Owen was separated from his herd. Just a juvenile and with no other hippos at the park, Owen formed an odd bond with Mzee (Swahili for "old man"), an Aldabra giant tortoise, that lasted until 2006. To Owen, Mzee resembled an adult hippo, with his large domed shell and brown color. By 2006, it became clear that Owen had grown too large; thus for Mzee's safety workers built a separate enclosure for Owen and bonded him with a new hippo named Cleo. Mzee was returned to his enclosure and these two famous friends went their separate ways.

Owney (Dog)

Owney, a border terrier, was dubbed the first unofficial mascot of the Albany, New York, post office and became a nationwide mascot for 9 years (1888–1897). In his lifetime he visited 48 states and traveled over 140,000 miles worldwide. In Irene Kelly's book *A Small Dog's Big Life: Around the World with Owney* (New York: Holiday House, 2005), the author invented letters, postcards and newspaper clippings based on fact to tell his story. To back up Owney's story clerks all over the United States had tags made to commemorate the dog's visit to their town. These tags are on view at the National Postal Museum, part of the Smithsonian Institution. The author admits taking liberties since there are many discrepancies in Owney's story, such as how he got his name. Owney was euthanized on June 11, 1897, at the age of 17. It seems that he bit a postal worker who was handling him roughly in front of a newspaper reporter. In any case it seems that Owney had a very interesting and adventurous life. James E. White,

Owney was the first unofficial mascot of the Albany, New York, post office, becoming a nationwide mascot between 1888 and 1897 (National Postal Museum Accession number: A.2008-42).

superintendent of the Railway Mail Service, collected money to have Owney preserved by a taxidermist; he is on display at the National Postal Museum.

Two children's books featuring Owney's story:

Owney the Mail-Pouch Pooch, by Mona Kerby, illustrations by Lynne Barasch. New York: Frances Foster Books, Farrar, Straus and Giroux, 2008

A Lucky Dog: Owney, U.S. Rail Mail Mascot, by Dirk Wales, illustrations by Diane Kenna. Chicago: Great Plains Press, 2003

Paddy (Dog)

Paddy, an Irish terrier belonging to Sergeant Major Beaumont Woodhead, was introduced as the mascot to a territorial unit: the 7th (Wellington West Coast) Regiment, based in Wanganui, New Zealand. In October 1914, Paddy was smuggled on board the troopship *Maunganui*

sailing to Egypt. The ship prohibited pets, but the men argued that based on Paddy's previous service, he was their unofficial mascot. After some training, Paddy accompanied the Australian and New Zealand Army Corps (the ANZACs), and on April 25, 1915, the battalion landed in Gallipoli under fire. Paddy participated during the Gallipoli campaign at the Battle of Chunuk Bair, where 77 men were killed and many more wounded. Due to the gunfire and shelling, Paddy became almost deaf and his nerves were greatly affected. As he recovered he went to live behind the lines with the quartermaster's staff. For his "good services and behavior" at Suez, Paddy was promoted to lance corporal, and by the end of the war he was a sergeant major. Paddy never made it back to New Zealand, and due to a rabies scare in Britain, Paddy lived at the Royal Society for the Prevention of Cruelty to Animals in London. Eventually he went to Devon to live with Jennetta Campbell Rogers. Paddy died in 1929 and Jennetta sent his collar to

Paddy, pictured here June 30, 1918, was the mascot of a territorial unit, the 7th (Wellington West Coast) Regiment based in Wanganui, joining the regiment in October 1914 (NZ National Army Museum).

the Wellington Regiment along with a condolence letter, which read in part, "I should like the Regiment to know that Paddy's last years (he must have been quite 16 when he died a short time ago) were as happy as care and kindness could make them…. He was a well-known character, and his war service made him an heroic figure."

Paddy (Dog)

The story of Paddy, a messenger dog, is both sad and remarkable. On both his first and second missions he was gassed, and on each occasion he somehow made his way back to headquarters. He was found completely blind in his kennel. He was hospitalized and made a full recovery both times. While on another mission he

Paddy, a messenger dog during World War I (photograph by Tom Aitken, National Library of Scotland, Creative Commons Attribution 4.0 International License).

was shot by a German and again somehow made it back to headquarters. Nothing is known of Paddy's final fate.

Paddy (Dog)

Handler MacLeod recalls the story of Paddy the Irish terrier, whose loyalty was tested when he managed to deliver his message to headquarters despite being gassed, wounded and left for dead.

Pal *see* Gander

Pal (Dog)

Pal, a male rough collie, was born on June 4, 1940, at Cherry Osborne's Glamis Kennels in North Hollywood, California. Pal is well-known as the first dog portraying the fictional female collie Lassie in film and television. MGM first hired Pal to be a stunt dog in its first "Lassie" film, with a female prize-winning show-collie to play the title character. Director Fred M. Wilcox decided to shoot some footage of the flooding of the San Joaquin River in Central California. With the female collie refusing to enter the water, animal trainer Rudd Weatherwax saw an opportunity to slip Pal into the shot. Pal's performance was so exceptional that the scene was completed in one take, and Wilcox was so impressed that the producers hired Pal for the role of Lassie. Pal's success in the 1943 film *Lassie Come Home* led to six more MGM films: *Son of Lassie, Courage of Lassie, Hills of Home, The Sun Comes Up, Challenge to Lassie,* and *The Painted Hills.* After the last film in 1951, MGM executives felt that the Lassie character was fading in popularity, and, planning no more films, they gave Pal's trainer Weatherwax the Lassie name and trademark.

Pal was the first dog to portray the fictional female collie Lassie in film and television.

Television producer Robert Maxwell and Weatherwax created the series *Lassie*, about the adventures of a boy and his dog on a farm in Middle America. Pal filmed the two pilots and CBS picked up the series for fall 1954. Pal retired after filming the two pilots and his son, Lassie Junior, took over the role. Pal died in 1958 of natural causes at age 18, leaving Weatherwax devastated. In 1950, Weatherwax and John H. Rothwell co-wrote a book about Pal's life, *The Story of Lassie: His Discovery and Training from Puppyhood to Stardom*. In 1992, *The Saturday Evening Post* wrote that Pal had "the most spectacular canine career in film history."

Pal (Dog)

Pal was killed in action at San Benedetto Po, Italy, on April 23, 1945. He bravely blocked a shrapnel charge with his body, preventing serious wounding of several men on advance patrol.

Pangur Ban (Cat)

Back in the ninth century, long before most people in Europe would have considered cats to be companion animals, an Irish monk, whose name is unknown, wrote a poem in Old Irish called "Pangur Bán" (Bán means "white" in Irish) about his relationship with his white cat by the same name. The poem demonstrates his admiration and love for his cat. It bears similarities to the poetry of Sedulius Scottus. In eight verses of four lines each, the author compares the cat's happy hunting with his own scholarly pursuits.

The poem about Pangur Bán is the earliest reference to a domesticated cat in European history:

> I and white Pangur
> practise each of us his special art:
> his mind is set on hunting,
> my mind on my special craft.
>
> I love (it is better than all fame) to be
>   quiet

> beside my book, diligently pursuing
>   knowledge.
> White Pangur does not envy me:
> he loves his childish craft.
>
> When the two of us (this tale never
>   wearies us) are
> alone together in our house,
> we have something to which we may
>   apply our skill,
> an endless sport.
>
> It is usual, at times, as a result of
>   warlike battlings,
> for a mouse to stick in his net.
> For my part, into my net
> falls some difficult rule of hard
>   meaning.
>
> He directs his bright eye
> against an enclosing wall.
> Though my clear eye is very weak
> I direct it against keenness of
>   knowledge.
>
> He is joyful with swift movement
> when a mouse sticks in his sharp paw.
> I too am joyful
> when I understand a dearly loved
>   difficult problem.
>
> Though we be thus at any time,
> neither of us hinders the other:
> each of us likes his craft,
> severally rejoicing in them.
>
> He it is who is master for himself
> of the work which he does every day.
> I can perform my own work
> directed at understanding clearly
>   what is difficult.

Translation by Gerard Murphy, *Early Irish Lyrics* (Oxford University Press, 1956), p. 3

Pauline Wayne (Cow)

Pauline Wayne was a 1,500-pound Holstein cow that captured both America's heart and the media's attention. In the 19th century there were no zoning ordinances against keeping cows; therefore when President William Howard Taft took office in 1909 there was nothing unusual about his cow, Mooly Wooly, supplying the White House with fresh milk and butter. Mooly Wooly suddenly died in 1910 and

This 1,500-pound Holstein cow (pictured in 1909) was a gift from Senator Isaac Stephenson of Wisconsin to President William Howard Taft (Library of Congress).

Wisconsin Senator Isaac Stephenson, seeing an opportunity for a headline, sent the Tafts a prize four-year-old Holstein. Pauline Wayne became a media sensation and her arrival in Washington was covered by *The New York Times*. Taft allowed Americans a chance to see his beloved pet in person at the 1911 International Dairyman's Exposition in Milwaukee. The White House came to a frantic standstill when Pauline disappeared along the way to the expo. Fortunately Pauline was spotted— her special car had gotten attached to a group of cattle cars headed for the Chicago Stockyards. With her health declining, in February 1913 she was sent back to Senator Stephenson's farm. Pauline's claim to fame was that she was the last cow to live on the White House premises.

Peefke (Dog)

Peefke's tour of duty ended when he was killed by an enemy hand grenade on March 20, 1945. On this mission, before his death, he had alerted his handler of a wire leading to three enemy "S" mines which were then neutralized, saving the patrol serious damage.

Peerless Pilot (Pigeon)

During World War I, Peerless Pilot became one of the most celebrated naval messengers. Beginning at only 15 months old, Peerless delivered nearly 200 messages from naval ships during the last year of the war. The U.S. Navy had over 1,500 pigeons at twelve stations in France. One of those stations, Pauillac, was home to Peerless Pilot. Navy aviators would take pigeons with them in their planes and release them in-flight for increased speed. The pilots were trained on how to release the pigeons so they would not be harmed.

Peggy (Dog)

When Peggy, an old English bulldog, was a puppy her owner donated her to the crew of the HMS *Iron Duke*. She was in a number of battles including the famous Battle of Jutland (Denmark) on May 31–June 1, 1916. The battle included the Royal Australian Navy and the Royal Canadian Navy against the Imperial German Navy's High Seas Fleet. Peggy was not the only dog on the ship (however her companion, Jumbo, also an English bulldog, had been left at the base because he got seasick). The couple had a litter of five pups. Peggy was known for being very playful and picked up stuff on the deck, for example hats and socks, and hid them in her kennel. Peggy would also hide under a gun turret, rush out, knock a sailor down, and then run and hide under another turret to wait for the next sailor. She loved to watch the men play soccer and sometimes rushed in to join them. The Royal Navy had a medal made for Peggy commemorating her participation in the Battle of Jutland. She also wore the General Service and Victory ribbons. After the war, she was returned to her owner and spent a few years raising money for St. Bartholomew's Hospital. In 1920, she was given back to the *Iron Duke* because it was clear that was where she really belonged. A destroyer made a special trip just to bring Peggy back to the ship, which was out to sea.

Peggy (Dog)

Peggy, a mixed breed, was awarded an inscribed collar by the RSPCA during World War II in England. When her house was bombed, her mistress was trapped under debris and a baby in a carriage was almost suffocated by falling plaster. Working with her paws, Peggy made a hole big enough for the baby to breathe through. All three—Peggy, the mother, and baby—were saved.

Pelorus Jack (Dolphin)

Pelorus Jack (sex never known)—13 feet long, white with grey lines or shadings, and a round, white head—was identified from photographs as a Risso's dolphin (*Grampus griseus*), rarely seen in New Zealand waters. Pelorus Jack was first seen in 1888 by the schooner *Brindle* and then became famous for escorting ships through Cook Strait, New Zealand. Pelorus Jack swam in Admiralty Bay between Cape Francis and Collinet Point, near French Pass, guiding ships through the dangerous channel between Wellington and Nelson. Jack would swim alongside the ships for twenty minutes at a time through dangerous waters, and there were no shipwrecks on Jack's watch. In 1904, someone aboard the SS *Penguin* attempted to shoot Jack with a rifle. Following the incident, on September 26, 1904, Jack was protected by Order in Council under the Sea Fisheries Act; he disappeared in 1912.

Penny (Leopard)

Having raised Elsa the lioness and Pippa the cheetah, Joy Adamson wanted to compare the three great cats of Africa. She was able to acquire a female orphaned leopard cub, Penny, in 1976. In the book *Queen of Shaba*, completed shortly before her death, Adamson described Penny's development from a pet to a wild leopardess. The cub had been young enough for Penny to become imprinted on Adamson—as the author described it, to "be given all the trust and affection that it would normally share only with its natural family." Adamson devoted three years to raising and rehabilitating this female leopard. Penny was a much more private, uncommunicative feline animal than Elsa. Penny was a biter and Adamson was always bandaging her arm after being nipped. Penny learned to catch and guard her prey, found a mate and gave birth to cubs. There were moments of affection, such as when Penny insisted that Adamson see her newborn cubs.

Pep (Dog)

Governor of Pennsylvania Gifford Pinchot received Pep as a gift in 1923. Pep was a beloved pet, but he developed a habit of chewing on the front porch sofa cushions, so the governor thought it best that he live somewhere else. In August 1924, Pep started his new life at the Eastern State Penitentiary in Philadelphia as a therapy dog for the inmates. He became famous when newspapers published that the governor sentenced Pep to life in prison

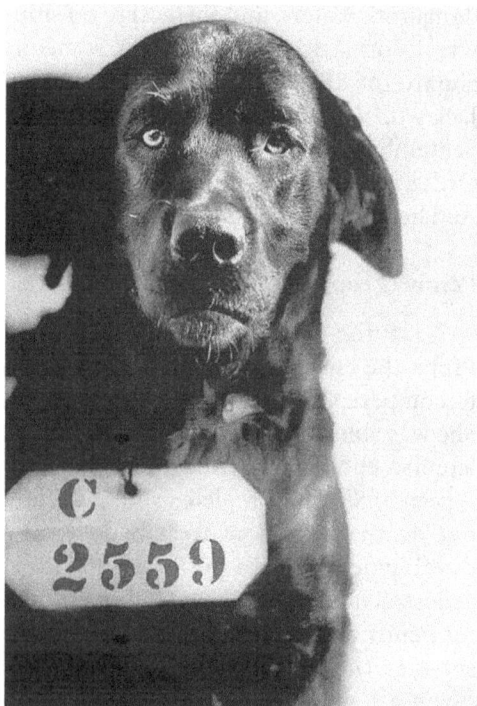

Pep, pictured here on August 31, 1924, was a therapy dog for the inmates at the Eastern State Penitentiary in Maine and is at the center of a humorous story now told at the penitentiary museum (Eastern State Penitentiary Museum).

without parole for killing his wife's cat, which turned out to be a false accusation. For a joke, the guards took the dog's paw prints and a mugshot with an inmate number. In 1926, the governor's wife, Cornelia Bryce-Pinchot, cleared Pep's name with a statement to *The New York Times*. In 1929, Pep was transferred to Graterford Prison, where he eventually died of natural causes; he is buried on the prison grounds. In reality, Pep was never locked in a cell and was free to wander around the prison. This humorous story is told at the Eastern State Penitentiary Museum; the prison closed in 1971.

Pete (Squirrel)

President Warren G. Harding and his wife Florence were great animal lovers. Pete the squirrel lived on the grounds of the White House and was seen running through the halls. Pete attended press conferences and news briefings. Pete would even eat right out of people's hands. In 1922, a correspondent reported that Pete was one of the last to leave a meeting. It was also reported that Pete was actually a mother of a litter of kits living in a nest on the grounds. It is thought that after Harding's death, Pete returned to living outside on the grounds. Harding was not the only president to have a Pete—President Harry Truman also had a pet squirrel named Pete.

Phar Lap (Horse)

Phar Lap was a chestnut gelding foaled on October 4, 1926, in Seadown, New Zealand. In 1928, businessman David Davis bought the horse as a yearling, and he was disappointed that the horse was ill-proportioned, awkward and sometimes described as a "donkey," a "throwback to a giraffe" and a "milk cart nag." Harry Telford, a trainer from Sydney, seeing some potential in Phar Lap, leased him for three

Phar Lap, shown here around 1930, is famous for his amazing career as a racehorse, winning 37 of 51 races (State Library, Victoria collection).

years and was not sorry. Phar Lap's name comes from the common Zhuang and Thai words for lightning, "sky flash" in English. Other names for Phar Lap were "Wonder Horse," "Red Terror," "Bobby," "Big Red," and sometimes "Australia's Wonder Horse." Phar Lap turned out to have a brilliant career as a racehorse, winning 37 of 51 races he entered. On April 27, 1929, he won his first race, the Maiden Juvenile Handicap at Rosehill, ridden by a 17-year-old apprentice, Jack Baker. On November 1, 1930, Phar Lap was shot at by criminals who wanted to stop his successes, but they missed and later that day he won the Melbourne Stakes and three days later the Melbourne Cup. In 1930, this racehorse demonstrated his incredible stamina: on a Saturday he won the Melbourne Stakes; the following Tuesday the Melbourne Cup; on Thursday the Linlithgow Stakes; and the

following Saturday he won the Fisher Plate. Phar Lap won his last race in Mexico in the Agua Caliente Handicap in track-record time. Phar Lap died on April 5, 1932, under suspicious circumstances, spawning many theories including accidental or purposeful arsenic poisoning, a stomach condition, and a bacterial infection. After six hairs from his mane had been analyzed, on June 19, 1908, the Melbourne Museum released the forensic findings of Dr. Ivan Kempson from the University of South Australia, and Dermot Henry from the Natural Science Collection at Museums Victoria. They deduced that in the 30 to 40 hours before his death Phar Lap ingested a massive amount of arsenic. It was concluded that the source of the massive amount of arsenic in Phar Lap's system would most likely never be determined. Contradicting the 1908 findings, *The Sydney Morning Herald* published an article in October 2011 claiming that Phar Lap had died of an infection, not arsenic poisoning. Phar Lap's heart was donated to the Institute of Anatomy in Canberra, his skeleton to the Te Papa Tongarewa museum in Wellington, and his stuffed body to the Melbourne Museum. In September 2010, both his hide and skeleton were displayed at the Melbourne Museum for celebration of the 150th running of the Melbourne Cup. Peter Luck's 1979 television series *This Fabulous Century* claims that Phar Lap's heart is really the heart of a draught horse. There is a 1983 film, *Phar Lap*, and a song, "Phar Lap—Farewell to You," honoring the horse. In both the Australian Racing Hall of Fame and New Zealand Racing Hall of Fame, he is one of five inaugural inductees. Phar Lap was ranked number 22 in *Blood-Horse* magazine's top 100 Thoroughbred champions of the 20th century. In 1978, the Australia Post placed his image on a postage stamp. Phar Lap has two residential streets named after him—in Bossley Park in Sydney, Australia, and in Cupertino, California. There is a Phar Lap Grove near the Trentham Racecourse in Wellington, New Zealand. Truly showing what a national icon Phar Lap is, there are two memorials dedicated to him: on November 25, 2009, a $500,000 life-size bronze memorial was unveiled near his birthplace at Timaru; also there is life-sized bronze statue at Flemington Racecourse in Melbourne. Phar Lap truly had an incredible life for a horse that started out having none of a Thoroughbred's traditional sleek lines.

Philly (Dog)

Philly was a stray mixed-breed female dog who served on the front lines of World War I with Company A of the 315th Infantry, 79th Division. On September 27, 1917, she was found by a soldier returning from Camp Meade and became a mascot. She was smuggled on board at the time of the unit's deployment to France. Throughout her service at Montfaucon, Montlouis, Troyon, and La Grande Montagne, Philly was a capable guard dog; her warnings of German sneak attacks resulted in a bounty of 50 deutsche marks for her death. She was wounded and gassed and still gave birth to four puppies. She was hailed a hero by the American soldiers. After the war, Philly marched wearing a blue army blanket with the 315th Infantry in their victory parade. In 1932, Philly died at the age of 15. Her body ultimately ended up at the Philadelphia History Museum.

Phoebe (Tiger)

Roman Emperor Nero first saw his pet tigress, Phoebe, fighting in the Coliseum with more ferocity than three tigers combined and decided to keep her as his companion. His servants built a golden cage on the palace grounds but did not keep her locked up all the time. Phoebe would share dinner at the emperor's table, even with guests, who could end up as Phoebe's dessert should anyone annoy or irritate Nero.

Pickles (Dog)

Pickles, a black and white collie, was referred to as the hero dog that found the stolen World Cup trophy just four months before the 1966 FIFA World Cup matches. Pickles saved England from international embarrassment when one day he was out walking with his owner, David Corbett, and he sniffed something in the bushes, which happened to be the World Cup trophy. He was lavished with a banquet in his honor and awarded with a bone and a check for 5,000 lira. In 1967, Pickles was awarded the silver medal of the National Canine Defence League. In the 1966 film *The Spy with a Cold Nose*, Pickles starred with Eric Sykes and June Whitfield. He also appeared in the television programs *Blue Peter* and *Magpie*. He was named "Dog of the Year" and pet food manufacturer Spillers awarded him a year of pet food. Pickles died in 1967 while chasing a cat; his choke chain caught on a tree branch and he was strangled. An ITV drama in 2006, *The Dog Who Won the World Cup*, written by Michael Chaplin, was entirely fiction. Pickles is buried in Corbett's back garden and his collar is on display at the National Football Museum in Manchester. In December 1983, the cup was stolen again and never recovered.

Pilot (Dog)

Reportedly George Washington had over fifty dogs during his lifetime, and it is believed that his dog Pilot (a "water dog," another name for poodle) was the first presidential poodle. In her research, author Kate Kelly connected with Emily Cain, who ran the "Poodle History Project," which had comprehensive information on poodles. Cain referred Kelly to a page on the project's website dedicated to *Companions to Genius* (the site is no longer active). On this page there were direct quotes from Washington's diaries about Pilot, who accompanied the president on many duck hunts.

"George Washington's Poodle" by Kate Kelly, Americacomesalive.com.

Pilot's Luck (Pigeon)

This remarkable bird has three amazing stories to his name. In the first account, Pilot's Luck was released from a seaplane 200 miles from his loft. When the plane had engine trouble and was attacked by the enemy, Pilot flew the 200 miles in just five hours. Then, he and another bird rescued a seaplane crew by bringing home a message for help. In a third account, he rescued a crew by flying fifty miles with a message.

Pinto (Horse)

The "Overland Westerners" were five men led by George Beck from Shelton, Washington, who began their journey in 1912 to ride horseback to every state capital in the United States. The group's goal was to gain fame and fortune by finishing their journey at the 1915 World's Fair in San Francisco. Pinto was Beck's horse, the only horse to date to walk 20,352 miles throughout the continental United States. The group's plan did not quite work out the way they had hoped. During their three-year journey they were broke, sore, cold and hungry. When the group arrived in San Francisco on June 1, 1915, their dreams were dashed when nobody would write about their story. Fast forward 100 years and Pinto "The Wonder Horse" would be an instant celebrity on websites, blogs, videos, and more.

Pipaluk (Polar Bear)

Pipaluk, a male, was born on December 1, 1967, at the London Zoo (following the only other polar bear cub reared at the zoo, Brumas, a female). Pipaluk's parents, Sam and Sally, came to the zoo as young cubs in 1960 from a Moscow zoo. The name Pipaluk means "the little one" in Inuit. In 1985, Pipaluk left the London Zoo when the Mappin Terraces, the enclosure housing all the bears, closed. Pipaluk died in a Poland zoo at age twenty-two.

Pippa (Cheetah)

Pippa had been raised as a pet by Major and Mrs. Dunkey until the age of eight months. They called her "Kitten" and wanted her to stay in Africa after they moved, so they gave her to Joy Adamson. Adamson renamed her Pippa and moved her to Kenya's Meru National Park. There Pippa developed her stalking and killing instincts and also mated. She was an aloof animal who lived in the bush several miles from camp but came in regularly for food during her pregnancies. Pippa would reappear to guide Adamson to see her new family: "Her trust in showing me her new family while their eyes were still closed moved me deeply," Adamson wrote in a book about Pippa's return to the wild, *The Spotted Sphinx* (1969). Pippa had four litters before she was injured in a fight and died. In Adamson's autobiography, *The Searching Spirit* (1978), she reported that on a visit to Meru Park in 1976, she saw one of Pippa's cubs, nine-year-old Whitey, playing with her cubs.

Pippin (Deer)

When Audrey Hepburn was making her 1959 film *Green Mansions*, the animal trainer suggested that she take her on-screen sidekick, a fawn, home with her to bond. Pippin thought of Hepburn as her mother. She nicknamed the fawn "Ip"; the deer would cuddle up with her and accompanied her to the market. The fawn slept in a custom-made bathtub. Hepburn's Yorkshire terrier, Mr. Famous, was very jealous of the new occupant.

Pitoutchi (Cat)

During World War I, Pitoutchi was one of eight kittens born in the Belgian trenches. Their mother was killed before their eyes even opened. Lieutenant Lekeux was on duty and though he tried, he was only able to save the one kitten. Pitoutchi never grew very big but was very intelligent. One day, the soldier encountered German troops on patrol. He hoped the Germans wouldn't see him hiding in the shell hole, but he heard one say, "He's in the hole." Just then Pitoutchi jumped from the hole. The Germans shot at the cat, missed and laughed that they had mistaken a cat for the enemy.

In the Belgian military records is the following commendation:

Pitoutchi, 3rd Regiment of Artillery, for showing great bravery under fire, rare endurance, and remarkable initiative. Showed proof, in the course of a campaign, of the finest military qualities. Seeing his captain in danger, did not hesitate to expose himself in his place, courageously drawing upon himself the enemy fire, and foiling the maneuvers of the adversary by making them mistake the above-mentioned officer for a cat. At the front since his birth. (Baker, *Animal War Heroes*, 69.)

Poll (Parrot)

Today many people own pet parrots for their intelligence, bright colors, and ability to mimic speech. President Andrew Jackson bought Poll, an African grey parrot, as a gift for his wife, Rachel, who unfortunately died, leaving the president to care for the bird. When the president died, Poll attended the funeral but had to be removed from the ceremony because he began swearing and yelling blasphemies in both English and Spanish. It turned out he had learned them from the president!

Pollux *see* Castor

Polo (Dog)

Polo is an example of how sentry dogs warned their handlers about German patrols in the area who were trying to capture soldiers on sentry duty. To warn his handler, Polo would point in the direction and then growl.

Pompey (Dog)

Pompey, a pug, was credited with saving his master, William I, Prince of Orange (AKA William the Silent), from an assassination attempt while he was asleep in his tent. Pompey's persistent barking and scratching and a leap onto his master's face woke the prince, foiling the assassination.

In 1584, in Pompey's absence, William was assassinated. William III of Orange, William's grandson, brought pugs along with him when he went to England to join his wife, Mary II, on the throne.

Poplar *see* Blackwall

President Wilson (Pigeon)

Amazingly this famous pigeon managed to fly twelve and a half miles in twenty-one minutes through rain and fog, despite having only one leg.

Prince (Dog)

A common story during World War I depicted dogs running away from home and finding their masters in the trenches. Most of these tales were fiction or wishful thinking. However the story of Prince, an Irish terrier, has been proven to be true. When Mr. Brown left for France, Prince was so upset that when one day Mrs. Brown let him out, he was not seen again—that is until he showed up in Private Brown's trench in France. Word spread about Prince's appearance and he was adopted as the regiment's mascot and stayed with Mr. Brown for the entire war. The soldiers taught Prince tricks and he was a source of amusement and comfort to the troops. He also learned to hide when he heard incoming heavy shells. Prince earned his keep by killing the rats in the trenches (137 rats in one day). One day Prince disappeared and was assumed dead, but a message came from a unit ten miles away that it had Prince. He returned to England after the war with help from the RSPCA (Royal Society for the Prevention of Cruelty to Animals). Prince was so popular and well-known that people sent cards and visited him.

Prince (Horse)

In 1878, Prince, a strawberry roan, and his owner, Chester Evans, were racing with 15 Cheyenne members to their destination at Fort Monument on Smoky Hill River in Kansas. Despite recovering from their wounds from arrows, the duo seemed to be greatly affected by the experience, especially Prince. Prince's lasting fear was demonstrated twenty years later when he was peacefully grazing in a pasture. Startling Evans, Prince raised his head, sniffed the wind and furiously began snorting and pawing at the ground as if to challenge an enemy. It was later that Evans learned that a group of Oklahoma Native Americans had stopped in Lebo, a small town half a mile from the pasture.

Prusco (Dog)

Prusco was a French wolfhound mix who dragged to safety more than 100 wounded soldiers who had fallen in brush, depressions and craters during battle in WWI.

Psyche *see* Irma

"The Purpose to My Life" (Pigeon)

Nikola Tesla had many pigeons he fed and cared for, but one he was particularly fond of. In 1924, this beautiful female pigeon flew into Tesla's window and died right before his eyes. Tesla later claimed a dazzling light had emerged from her eyes while dying. After that experience, Tesla was never the same and was never as creative. If this phenomenon hadn't occurred, he may have been able to create more inventions.

Pushinka (Dog)

Pushinka (whose name is Russian for "fluffy") is famous in history because she was a gift from Soviet Premier Nikita Khrushchev to President Kennedy's daughter Caroline in the midst of the Cold War; therefore she was x-rayed and thoroughly

Pushinka, pictured here June 22, 1961, with the Kennedys' Welsh terrier Charlie, was a gift from Soviet premier Nikita Khruschev to President Kennedy's daughter Caroline (Robert Knudsen. White House Photographs. John F. Kennedy Presidential Library and Museum, Boston).

searched for "bugs" or a possible dooms-day device before being allowed in the presidential quarters. She was also famous for being a mixed-breed puppy sired by Strelka, a famous Soviet space dog. He made history for being one of the first living creatures to orbit Earth aboard *Sputnik* 2 and returning unharmed on August 19, 1960. In June 1963, Pushinka and Charlie, the Kennedy's Welsh terrier, had a litter of four puppies named Butterfly, White Tips, Blackie and Streaker. President Kennedy referred to them as the "pupniks." The White House held a contest by selecting ten finalist letters from 5,000 that had been sent asking for a puppy. Jackie Kennedy chose two children to win a puppy. The lucky children were ten-year-old Karen House from Illinois (who got Butterfly) and nine-year-old Mark Bruce from Missouri (who got Streaker). The two remaining puppies were given to family friends.

Pyrame (Dog)

Pyrame's barking signaled the presence of German soldiers, thus saving a battalion of soldiers.

Rachel Alexandra (Horse)

Rachel Alexandra, a bay mare foaled in Kentucky on January 29, 2006, was sired by Madaglia d'Oro, and her dam was Lotta Kim. (Madaglia d'Oro was foaled 1999 in Kentucky and had a very successful racing career, winning multiple Grade I stakes and earning $5,754,720.) Rachel Alexandra was shunned by her mother; the foal was scrawny and had a medical condition that affected her bones and joints. Despite most people counting her out of any kind of a racing career, she proved to be a great athlete, breaking numerous records. Ridden by jockey Calvin Borel, Rachel Alexandra became the first filly to win the Preakness in 85 years and also the first filly to win the Woodward Stakes. Her total earnings were $3,506,730, and some of her honors and awards included American Champion Three-Year-Old Filly (2009) and American Horse of the Year (2009). On January 18, 2010, she won the Eclipse Award for Champion Three-Year-Old Filly, and on September 29, 2010, The Fairgrounds Race Course renamed the Silverbulletday Stakes the Rachel Alexandra Stakes to honor her. On April 25, 2016, she was inducted into the National Museum of Racing and Hall of Fame. On March 25, 2012, Rachel Alexander was bred with Bernardini at Stonestreet Farm, and their foal, named Rachel's Valentina, was the offspring of two Preakness winners. Deb Aronson's biography "shows how truly exciting the world of horse racing can be, using the story of the record-breaking filly who beat all the boys"—Blake Norby.

Rags (Dog)

Rags' story began when he was found abandoned in Paris by Private James Donovan, an A.E.F. signal corps specialist, and George Hickman, while serving with the U.S. 1st Infantry Division in World War I. Donovan appropriately named this mixed-breed terrier Rags because he mistakenly took him for a pile of rags. Rags became the mascot of the 1st Infantry Division, resulting when Donovan was late reporting back to his unit after marching in the Bastille Day Parade. To avoid being Absent Without Leave, he told the military police that he was part of a search party for the 1st Infantry Division's missing mascot. In addition to being a mascot,

After he was found abandoned in Paris, Rags became the mascot of the U.S. 1st Infantry Division in World War I. Rags is shown in this undated photo with Sergeant George E. Hickman, 16th Infantry, 26th Division (U.S. Army Signal Corps).

Rags was also useful as an early-warning signal. Rags could hear an incoming mortar attack before the men, so they learned that whenever Rags flattened himself on the ground, they should also do so. Donovan's jobs were to string communications wire between advancing infantry, support field artillery, and also repair field telephone wires. When Rags worked alongside Donovan, the dog could spot open wires and mark them so Donovan was able to make the repairs. Until the work was completed runners were used, but they were often wounded, killed or could not get through with their messages. Therefore Donovan trained Rags to carry messages in his collar. Of note, in July 1918 Donovan, Rags, and an infantry unit of 42 men were surrounded by Germans. Rags carried back a message prompting an artillery barrage and reinforcements that saved the group. Also of note, in the final American campaign, and among the many messages that Rags had carried, was a very important message on October 2, 1918. Rags ran it from the 1st Battalion of the 26th Infantry Regiment to the 7th Field Artillery, leading to the Very-Epinonville Road being secured, saving many lives. Sadly, on October 9, 1918, Rags and Donovan were hit by German shellfire and gas shells in the final American battle in the Meuse-Argonne. Rags' right front paw was injured, and he lost hearing in his right ear, lost sight in his right eye and was mildly gassed. Donovan was seriously wounded and badly gassed. The two were taken to several hospitals and while Rags quickly healed Donovan got worse. They were returned to the United States, and in early 1919 Donovan died and Rags became the post dog, making his home at the base firehouse. A tag on his collar identified him as "1st Division Rags." In 1920, Rags made his new home with Major Raymond W. Hardenbergh, his wife, and two daughters. In 1924, Rags and the Hardenbergh arrived at Governors Island in New York Harbor, where Rags became a New York celebrity. He participated in parades and numerous articles about him appeared in *The New York Times*. Author Jack Rohan wrote a book about him published in 1930, and Rags was awarded a number of medals and awards. Also New York politicians and U.S. Army generals had their pictures taken with Rags. There is little known about Rags' life between 1934 and 1936, after Hardenbergh was promoted to lieutenant colonel and was transferred to serve in the War Department in Washington, D.C. In March 1936, Rags died at age twenty and was buried with military honors. A monument was erected at the Aspin Hill Memorial Park and pet cemetery in Silver Spring, Maryland.

Rajah (Dog)

Rajah, a German shepherd dog, was unofficially recognized as the first police dog in New Zealand during the 1930s. His owner, Constable John Robertson, trained Rajah to locate hidden weapons and find bodies of missing persons. In addition to his police work, the dog also made public appearances to demonstrate his talent for finding hidden objects. A side note: After the original Rin Tin Tin died, Rajah was offered the role, but his family decided to keep him in New Zealand.

Rats (Dog)

Rats, the soldier dog from Ulster, was a little brown and white mongrel who served in Crossmaglen in Northern Ireland. He spent most of his service time with the troops on the front lines going out on day and night helicopter missions. Also in the line of duty he sustained many different injuries including a broken leg, various gunshot wounds, and shrapnel in his spine. Because Rats was also a mascot, morale booster, and so good at detecting potential danger, he became a target of the IRA. One of his most difficult times came when the troops he had become so attached to had come to the end of their service and a new regiment took their place. Corporal O'Neil of the Queen's Own Highlanders took care of Rats and the two became close friends. Once again Rats lost a good companion when O'Neil's regiment was replaced with the Prince of Wales Company, 1st Battalion, Welsh Guards. In time, Rats made new friends and became close to Corporal Lewis. Rats had a happy retirement in an unknown location in the UK to keep him safe from the IRA. Rats was awarded a medal with the queen's head on one side and on the other side his name and army number, "Rats. Delta 777." Author Max Halstock wrote a book, *Rats: The Story of a Dog Soldier*, published in 1981.

Ratty *see* **Commuting Animals of England**

Rebecca (Raccoon)

Rebecca was sent to President Calvin Coolidge in late November 1926 by Vinnie Joyce of Nitta Yuma, Mississippi, and was destined for the president's Thanksgiving dinner table. However, President Coolidge took a shine to the raccoon and granted her a pardon. Today the thought of eating raccoon for Thanksgiving dinner is very odd; however, the *Washington Evening Star* thought it strange the president did not want to eat this animal. The paper declared that raccoon meat is less fatty and rather tasty: "The ring-tailed animal tasted like chicken, albeit one crossed with a suckling pig" (History.com). It was not surprising that the president and first lady were well-known as animal lovers, and people began to send them unsolicited pets. Coolidge wrote in his autobiography, "We always had more dogs than we could take care of." While the canines, cats and canaries that were sent to the president may have been on the conventional side, Coolidge also received a black-haired bear from Mexico, an African pygmy hippopotamus from magnate Harvey Firestone, and even a pair of lion cubs, which the fiscally conservative president gave the less-than-fuzzy names Tax Reduction and Budget Bureau. Rebecca proved to be a very mischievous pet, ripping clothing, clawing the upholstery, wriggling out of harnesses, and escaping her makeshift cages—leading to chaos in the president's

In November 1926, Rebecca was sent from Nitta Yuma, Mississippi, by Vinnie Joyce to President Calvin Coolidge for his Thanksgiving dinner, but was granted a pardon after the president took a shine to her. Grace Coolidge holds Rebecca at the Easter egg rolling on April 18, 1927 (Library of Congress).

home. To remedy the situation, staff built a wooden house and placed it in a tree on the White House's South Lawn, where the president could see her through a window. For Christmas she was named Rebecca and given a shiny collar engraved, "Rebecca Raccoon of the White House." President Coolidge was so attached to Rebecca that he walked her on a leash and cuddled her in his lap in front of the fireplace. In March 1927, she even moved with the president into a Dupont Circle mansion while the White House was under renovation. She was once banished to the National Zoo when it appeared that she may have bitten the president (he was seen with his wrist bandaged), but she was out in less than a week. In the summer of 1927, Rebecca, five canaries, and two white collies, Rob Roy and Prudence Prim, accompanied the president on his three-month vacation to South Dakota's Black Hills. By early 1928, because she was escaping more and more to roam the capital city, the Coolidges donated her to the National Zoo. Today it is illegal to keep a pet raccoon in Washington, D.C.

Red Cock (Pigeon)

When Captain Crisp's trawler *Nelson* was attacked by a German U-boat, he was killed, but before he died he was able to send Red Cock (AKA Crisp V.C.) off with a message for help. Red Cock delivered the message to a nearby ship, helping to save the surviving crew. Red Cock was renamed Crisp V.C. for this act. He was honored at the United Services Museum in Whitehall, London, England.

Red Ghost (Camel)

The U.S. Army introduced camels to the Southwest back in the 1880s, using them as beasts of burden while surveying, or as a cavalry tool. But the Civil War interrupted the camel experiment and most were sold or released to run wild. The tale of Red Ghost starts in 1883 at a lone ranch in Arizona, where a woman's body was found trampled to death. The next day tracks were found, cloven hoof prints much larger than those of a horse, along with long strands of reddish hair. A few days later prospectors were awakened by thundering hooves and screams. They saw a gigantic creature running off! The next day they found huge hoof prints and red hair. Stories multiplied and were embellished by the locals. One man claimed to see Red Ghost kill and eat a grizzly bear. Months later the camel was spotted and shot at but missed. The animal bolted and a strange object fell to the ground. It was a human skull. The creature kept making appearances and one cowboy recognized the skeletal remains of a man attached to the camel's back. In 1893, a rancher brought down Red Ghost with a single shot. The origin of the skeletal rider has been lost to conjecture. The animal's back was scarred from rawhide strips that had been used to tie down the body of a man.

Red Rum (Horse)

Red Rum, a Thoroughbred bay gelding, won the Grand National in 1973, 1974, and 1977, and came in second in 1975 and 1976. He was laid to rest at the finishing post of Aintree Racecourse in England.

Red Rum's tombstone epitaph reads:

Respect this place / this hallowed ground / a legend here / his rest has found / his feet would fly / our spirits soar / he earned our love for evermore.

Regal (Horse)

Regal, a bay gelding, was one of 186 horses that were part of the Metropolitan Police Mounted Branch in London during World War II. These horses were used to aid civilians during the Blitz and throughout most of the war (from September 1940 to late 1944). Regal and his handler, P.C. Hector Poole, served in the north London suburb of Muswell Hill. On April 19, 1941, incendiary bombs were dropped, causing a fire to spread around Regal's stall, but he escaped without injury. However three years later, on July 20, 1944, he was not so lucky when a V-1 flying bomb caused the station roof to partially collapse, injuring Regal with flying debris. On April 11, 1947, Regal was awarded the Dickin Medal in a ceremony at Hyde Park.

Resi (Dog)

Resi was a female collie who served under the command of General von Blumenthal, a German officer, in 1899. She was trained to seek out missing and wounded soldiers in the battlefield when given the command "seek wounded." She was sent into a simulated battlefield with stretcher bearers to seek out missing wounded soldiers. Afterwards, there were still three soldiers missing and Resi was able to find them within ten minutes. Resi would also guard her trainer's gear while he tended to the wounded.

Rex (Dog)

In 2008, Rex, a mix between a German shorthaired pointer and a wire-haired pointer, and his owner, while on a walk came upon a dead kangaroo lying by the side of the road, and the two continued on home. A few hours later, Rex showed up at home with a joey (baby kangaroo) to everyone's surprise. He must have gone back to the body and rescued the joey. The joey was given the name Rex Jr. in honor of his rescuer. Rex Jr. was cared for at the Jirrahlinga Koala & Wildlife Sanctuary in Australia. When Rex Jr. was fully grown he was released back into the wild.

Rex (Horse)

Rex, a black Morgan stallion, starred in more than a dozen films over fifteen years despite his reputation for being "mean," "vicious," "ornery," "undependable," "warped," and "dangerous." Also known as "The Wonder Horse" and "King of the Wild Horses," he was the first horse to star in his own films. Because many actors refused to work with him, filmmakers often used a double in close-up shots. Some of Rex's silent and sound films include: *The King of the Wild Horses* (1924), *The Devil Horse* (1926), *No Man's Land* (1927), *King of the Wild Horses* (1933), *The Law of the Wild* (1934), *The Adventures of Rex and Rinty* (1935) and *King of the Sierras* (1938). It is somewhat puzzling that such a fierce horse as Rex often brought comedy and action into his films. For example, in one scene from *No Man's Land* Rex chases a pair of villains. He pushes one over a cliff and nudges the other one off a ledge into a watering hole—all to save the modesty of a young woman who is swimming in the nude.

Rienzi (Horse)

Rienzi (later known as Winchester) was a Morgan horse ridden by General Philip Sheridan during the Civil War. He was given to the general from the officers of Second Michigan Cavalry in 1862. His name was changed to Winchester after carrying Sheridan to victory on his ride from Winchester, Virginia, to Cedar Creek, Virginia, on October 19, 1864. Surviving the war, Rienzi lived to the age of 36. In 1879, his preserved and mounted body was presented to the Military Museum at Governors Island, New York, and remained there until a fire in 1922. After the fire, Rienzi was moved to the Smithsonian National Museum, and finally in 1964 he was moved to the National Museum of American History, where he wears General Sheridan's saddle and bridle. Rienzi and the general have been commemorated in poetry, songs, sculptures, museum exhibits, battlefield recollections, portraits, sketches and poems. The most famous poem, titled "Sheridan's Ride," was written by Thomas Buchanan Read. Rienzi was also famous as the subject of the painting *Sheridan's Ride*.

> Be it said in letters both bold and bright:
> "Here is the steed that saved the day
> By carrying Sheridan into the fight,
> From Winchester—twenty miles away!"

Rifleman Khan (Dog)

Rifleman Khan, a German shepherd dog, had been the family pet of the Railton family from Tolworth, Surrey. In the summer of 1942, the family lent Khan to the War Office to be a military dog during World War II. Khan and his handler, Lance Corporal James Muldoon, were assigned to the 6th Battalion of the Cameronians (Scottish Rifles). The island of Walcheren in the Netherlands needed to be taken for the invasion of Germany to take place. In November 1944, as part of the Battle of the Scheldt, Khan and Muldoon were approaching the island by sea in an assault craft when they came under heavy fire, capsizing the boat. Khan swam to shore, and when he did not see Muldoon he swam 200 yards through the sea and back, pulling Muldoon onto the shore. On March 27, 1945, Khan was awarded the Dickin Medal for bravery. His citation read: "For rescuing L/Cpl. Muldoon from drowning under heavy shell fire at the assault of Walcheren, November 1944, while serving with the 6th Cameronians (SR)." At a war dogs' parade at Wembley Stadium, Khan and Muldoon were reunited and spent their remaining years together in Strathaven.

Rin Tin Tin (Dog)

Rin Tin Tin was a male German shepherd that gained international fame in motion pictures. American Corporal Lee Duncan, Rin Tin Tin's trainer, rescued him as a puppy in a bombed-out kennel on a World War I battlefield in France and nicknamed him "Rinty." Duncan took on the training of Rin Tin Tin to be a star. The dog had his first break in the 1922 silent film *The Man From Hell's River*, where he replaced a camera-shy wolf not performing well. In another 1922 film, *My Dad*, Rin Tin Tin played a small role as a household dog. He picked up his first starring role in the 1923 film *Where the North Begins* with silent-film star Claire Adams. The huge success of this film is credited with saving Warner Bros. from bankruptcy. Rin Tin Tin went on to make 24 more films. For the first Academy Awards in 1929, he was nominated for best actor but lost to German actor Emil Jannings. Most of Rin Tin Tin's roles were in silent films; however, he did star in four "talkies," which included a 12-episode serial film, *The Lightning Warrior*, in 1931. By 1927 the dog was named the most popular performer in the U.S., and his films were seen in 70 countries. After Rin Tin Tin's death on August 10, 1932, regular programming was interrupted by a special news bulletin. Also after the celebrity dog's death, his name was given to several German shepherd dogs being featured on film, radio, and television. Duncan died in 1960 and after TV producer Herbert B. "Bert" Leonard inherited the screen property of Rin Tin Tin, he went on to make the Canadian TV show *Katts and Dog* from 1988–1993 (called *Rin Tin Tin: K-9 Cop* in the U.S. and *Rintintin Junior* in France). After Leonard's death in 2006 his attorney James Tierney made the 2007 film *Finding Rin Tin Tin*. In her second book, New York writer Susan Orlean not only wrote about Rin Tin Tin but also explored "the role of dogs in war, the evolution of Hollywood heroes, the nature of our attachment to our pets and the American faith in the qualities Orlean believes Rin Tin Tin embodied—strength, loyalty, devotion and truth."

Duncan's poem "Rin Tin Tin" reveals

Rin Tin Tin was the famous male German Shepherd making over 24 films and being nominated for best actor in the first Academy Awards in 1929.

his close relationship with the dog. The last four lines read:

> A real unselfish love like yours, old pal,
> Is something I shall never know again;
> And I must always be a better man,
> Because you loved me greatly, Rin Tin Tin.

Rip (Dog)

Rip is a mixed-breed terrier found as a stray by Air Raid Warden Mr. E. King after a bombing raid of Poplar, London in 1940. Rip was adopted as a mascot of the Southill Street Air Raid Patrol and became the service's first search and rescue dog, credited with saving over 100 victims of air raids in London. In 1945, he became the first of twelve dogs to be awarded the Dickin Medal for bravery, which is the animal equivalent of the Victoria Cross. The citation on the medal reads, "For locating many air-raid victims during the blitz of 1940." Rip died in 1946 and was buried in the PDSA Cemetery in Ilford, Essex. His headstone reads, "'Rip, D.M., We also serve'—for the dog whose body lies here played his part in the Battle of Britain." Rip was partially responsible for the training of search and rescue dogs at the end of World War II.

> How welcome to the victims must have been the first sounds of those scrabbling paws, shrill terrier yaps, and the first sight of the grinning Tommy Brock face with its merry friendly eyes.
> Jilly Cooper, *Animals in War*

Rob (Dog)

In 1942, the War Office appealed to the public to sign up their domesticated pets for training to fight in the war. From the thousands of animals signing up, only about 2,000 were sent for further training. One such dog was Rob Bayne, a black and white collie from the Bayne family farm in Shropshire. Listed as War Dog No. 471/322, Rob was assigned to the Special Air Service at the base in Wivenhoe Park, Essex. Rob's first duty was as a guard and messenger dog. On February 3, 1945, Rob received the Dickin Medal in London. The citation read, "For service including 20 parachute jumps while serving with Infantry in North Africa and SAS Regiment in Italy." However, Rob's parachuting exploits were revealed to be a hoax in an autobiography by SAS training officer Quentin "Jimmy" Hughes, titled *Who Cares Who Wins?* Tom Burt, the quartermaster for the 2nd SAS Unit, had grown attached to the dog, and when Rob's owners had requested to have him back, he and Hughes arranged to send Rob on a parachute jump; they had planned to then write to the family that Rob was indispensable. Their attempt on a parachute drop fell through due to high winds. Hughes still wrote to the owners, who were so proud of their dog that they passed the letter on to the PDSA; thus Rob was awarded the medal. In November 1945, Rob returned to the Baynes and he died on January 18, 1952. Rob was buried on the family farm, with a memorial stone inscribed:

> To the dear memory of Rob, war dog no 471/322, twice VC, Britain's first parachute dog, who served three and a half years in North Africa and Italy with the Second Special Air Service Regiment. Died 18th January 1952 aged 12½ years. Erected by Basil and Heather Bayne in memory of a faithful friend and playmate 1939–1952.

A children's book about Rob titled *Rob the Paradog*, written by Dorothy Nicolle, was published by Blue Hills Press.

Rob Roy (Dog)

Among Calvin Coolidge's menagerie of animals, some of which lived at the White House and others at the zoo, were his beloved Rob Roy and Prudence Prim, white collies. The other animals included a donkey named Ebenezer, a pygmy hippo named Billy, a wallaby, a bobcat, canaries, a pair of raccoons named Rebecca and Horace, and a pair of lions. Rob Roy came to live at the White House when Grace Coolidge fell in love with the breed after seeing them performing in a circus. Grace Coolidge purchased him from Stephen C. Radford Jr.'s Island White Kennels in Oshkosh, Wisconsin. An amusing bit of trivia was that Rob Roy's original name was Oshkosh, but the first lady renamed him after the popular cocktail Rob Roy, which is a bit amusing because it was the time of Prohibition. Rob Roy was the first dog included in an official First Family photo, a bit melancholy because the Coolidges' 16-year-old son Cal died a week later from an infected toe blister as a result of being in such a hurry to change clothes that he

Rob Roy, shown in a picture of September 26, 1924, was one of two white collies among Calvin Coolidge's menagerie of animals (Library of Congress).

office. His especial delight was to ride with me in the boats when I went fishing. So although I know he would bark for joy as the grim boatman ferried him across the dark waters of the Styx, yet his going left me lonely on the hither shore.

Rockie (Dog)

On November 19, 1997, Rockie was awarded the Purple Cross at Parliament House in Sydney for rescuing his owner from a fire at their home. Rockie, a four-year-old English bull terrier, smelled smoke while his owner was asleep. He took action by jumping at her bedroom door to wake her, and when she opened the door she was only semi-conscious from smoke inhalation. Without Rockie's lifesaving actions by dragging her body through the flames, she may have died within minutes.

Roger (Elephant)

The *Rocky Mountain News* account of the events occurring in July 1891:

Warning: the news story below is gruesome

> Terrorized Roger the big elephant at Manhattan Beach takes human life. Six-year-old George W. Eaton stamped out of existence. The brute was scared by the balloon which was about to ascend. Party of merry children were riding on the animal's broad back. The unfortunate child fell from the perch and was immediately trodden upon. One mass of blood and mangled flesh was found after it happened. Efforts made by the excited elephant's keeper to reach the lad. Hundreds of people witness the sickening sight and a panic is created.

The tragedy of Pressley Eaton's six-year-old son George W. Eaton, who was killed by Roger the elephant, took place at Manhattan Beach, the first amusement park west of the Mississippi River, built on the shore of Sloan's Lake in Edgewater, Colorado, and in operation from 1881–1914. It had many attractions such as a roller coaster,

Rob Roy pictured with President and Mrs. Coolidge ca. 1925.

forgot to wear socks. Rob Roy's legacy is depicted in a portrait by the American artist and illustrator Howard Chandler Christy of a seated Rob Roy gazing up at Grace Coolidge, who was wearing a striking red dress. This portrait still hangs in the China Room at the White House today. According to a 2007 book by Margaret Truman, *The President's House: 1800 to the Present* (Random House), the dog's gaze upward was achieved by feeding him candy from the artist's hand. Sadly Rob Roy died in September 1928, when he did not survive emergency surgery for stomach problems. In his autobiography, Calvin Coolidge wrote about his beloved pet:

> He was a stately companion of great courage and fidelity. He loved to bark from the second-story windows and around the South Grounds. Nights he remained in my room and afternoons went with me to the

hot air balloon rides, and contortionists, but the biggest draw was Roger the elephant (real name: Rajah), who gave children rides on his back. The chaos began a little after 4 o'clock in the afternoon when Roger, frightened by the ascension of a hot air balloon, broke away from the keeper, throwing the boy to the ground. Roger was put down and buried in a nearby swamp. In 1908, Manhattan Beach was destroyed by fire.

Rolf (Dog)

Rolf, an Airedale terrier, was one of Nazi Germany's super dogs. Rolf's owner, Paula Morkel from Mannheim, Germany, claimed that Rolf could communicate with his paws by tapping various times to represent the letters of the alphabet. She even claimed he was a poet and could speak several languages. Morkel's claims caught the attention of the Germans, whose animal psychologists viewed canines as being almost as intelligent as humans and thought they could be trained to reach their potential intelligence. In 1930 Margarethe Schmidt founded the Hundesprechschule Asra ("Asra talking school for dogs") to teach the animals to talk, count and reason. An article in *Psychology Today* concluded that Rolf proved to be influential in the development of the "Nazi talking dog program."

Roman (Dog)

Handler MacLeod recalls Roman, a purebred, tri-colored collie who appeared to be very self-centered during training but turned out to be very devoted to his job. He showed an ability to analyze a situation and determine right from wrong.

Room Eight (Cat)

Room Eight was a domestic short-haired cat who in 1952 climbed through a window in Room Eight at Elysian Heights Elementary School in Echo Park, California. The students named the cat Room Eight because for sixteen years he would return to the room every fall. When he was older he was injured in a cat fight and also suffered from feline pneumonia. He died on August 13, 1968, and the students attended his funeral at The Los Angeles Pet Memorial Park; his paw prints are memorialized in the school's cement sidewalks. He was featured in a documentary, *Big Cat, Little Cat*, and in a 1966 children's book, *A Cat Called Room 8* by Virginia Finley and Beverly Mason and illustrated by Valerie Martin (published by G.P. Putnam's Sons). In November 1962, *Look* magazine ran a three-page article titled "Room 8: The School Cat." In Leo Kottke's 1971 album *Mudlark* he wrote an instrumental piece titled "Room 8." In 1972, a cat shelter was named after the cat, "The Room 8 Foundation."

Rose (Orangutan)

Napoleon's wife Josephine kept a number of exotic pets, such as black swans, emus, and kangaroos, but her favorite was Rose, a female orangutan that was given to Josephine by General Charles Decaen, governor of the Île de France. Rose was remarkably well behaved, wearing a white chemise dress, rising to greet guests and having excellent table manners using a knife and fork. However, the orangutan loved turnips. Once the vegetable was served, she forgot all her manners and would run around and do somersaults—charming Josephine so that she thought of Rose less as an animal and more like an adorable misbehaved little girl. Rose died within a year of her arrival. After grieving, Josephine gave her body to the Museum of Natural History, and the museum's director, Frederic Cuvier, wrote an entire volume, *Description of an Orangutan and Observation of Its*

Intellectual Faculties, about his studies of Rose.

Rosell *see* **Apollo**

Rover *see* **Blair**

Roy (Dog)

Roy was a canine soldier that served in occupied Europe and the Far East to fight against Hirohito's forces—parachuting behind enemy lines during World War II from 1939 to 1945. He finished his career as a police patrol dog. Roy was a German shepherd given to Jim Love of Bathgate by his uncle John Arthur in 1939. When it began to get tough to feed both the family and a full-grown German shepherd on rations, the family donated Roy to the war effort. He trained at the Guard Dog Training School in Woodfield before moving to Europe. The service of this brave war dog was highlighted when Jim Love sought to get a copy of Roy's war certificate, which praised his "loyal and faithful service," to show his grandson for a school project on World War II. Roy lived to be 18 years old. Mr. Love said, "He was an inspiration during the dark years of the Second World War, serving his country as loyally as he was to me before and after."

Ruby (Elephant)

Ruby was an Asian elephant born in Thailand and bought by the Phoenix Zoo in 1974 at about seven months old. Noticing Ruby scratching in the dirt with a stick, her keepers jokingly gave her a paintbrush and some paints. Soon Ruby was creating modern art paintings. One of her paintings sold for as much as $25,000. The Phoenix Zoo was able to make many improvements from the money it made selling Ruby's art. In 1998, Ruby died of complications from pregnancy.

Rufus and Rufus II (Dogs)

Winston Churchill, an animal lover, owned two brown poodles named Rufus. In 1947, Rufus I was killed in a road accident. Churchill was so distraught over his loss that he got another dog with a strong resemblance to his beloved Rufus and named him Rufus II.

Rumpel (Cat)

Poet Laureate Robert Southey was a real cat lover, with his felines making frequent mention in his writings. In 1826, he wrote a letter to his seven-year-old son:

> I hope Rumpelstiltzchen has recovered his health, and that Miss Cat is well....

Seven years later, Rumpel's health finally gave out. Southey shared the news:

> Alas! Grosvenor, this day poor old Rumpel was found dead, after as long and happy a life....

Rumpel was formally named "The Most Noble the Archduke Rumpelstiltzchen, Marquis Macbum, Earle Tomemange, Baron Raticide, Waowler, and Skaratchi."

After Rumpel died, Southey wrote:

> We are each and all, servants included, more sorry for his loss, or rather, more affected by it, than any of us would like to confess.

Rusik (Cat)

Rusik was killed in the line of duty. His fate was sealed in 2002 when he wandered into the police station at their checkpoint on the Caspian Sea and replaced the resident dog and fish sniffer. Rusik became invaluable in sniffing out poachers of sturgeon, which has a dwindling population. Because of money to be made in the illicit trade of caviar, a mob contract was put out on Rusik and he was run over by a car in 2003.

Ruswarp (Dog)

Ruswarp, a very loyal 14-year-old border collie, was found guarding the dead body of his owner, Graham Nutall, 11 weeks after the dog and his owner went missing on their usual walk in the mountains. Ruswarp was very weak and died on the same day they were found. Nutall had been a member of a group called Friends of the Settle Carlisle Line, which fought against the closure of the railway station in Garsdale, England, which is where Ruswarp's statue stands today.

Saint Guinefort (Dog)

Saint Guinefort was a 13th-century French dog referred to locally as a folk saint after miracles were reported at his grave. The legend goes that Guinefort, a greyhound, belonged to a knight who lived in a castle near Lyon. One day the knight went hunting and left Guinefort to take care of his infant son. Upon his return the nursery was in chaos and his son was nowhere to be seen. When the dog greeted him with bloody jaws, the knight killed Guinefort, believing that he had devoured his son. Afterward he heard his son crying, turned the cot over and found his son safe, with a dead viper lying beside him. Realizing his mistake—that in fact Guinefort had killed the snake, saving the child—the family dropped Guinefort down a well and built a shrine to him by covering his body with stones and planting trees around it. The locals believed Guinefort to be a saint for the protection of infants, and allegedly they left their babies at the shrine to be healed by the dog. Despite prohibitions by the Catholic Church, the cult of this dog saint lasted for several centuries until the 1930s. There is a 1987 French film, *Le Moine et la Sorciere (The Sorceress)*, depicting the controversy over Saint Guinefort from the perspective of Father Étienne de Bourbon, a Dominican inquisitor. Also, in Bernard Cornwell's book trilogy *The Grail Quest* (2000–2003) the main character, Thomas of Hookton, is a mock believer who prays to Saint Guinefort wearing a paw on a piece of leather around his neck.

Sallie Ann Jarrett (Dog)

Sallie, a brindle bull terrier, became the mascot for the Eleventh Regiment of the Pennsylvania Infantry under the command of Colonel P. Jarrett, accompanying the soldiers during almost the entire Civil War. One day in May 1861 a resident of West Chester, Pennsylvania, gave this four-week-old puppy to 1st Lt. William R. Terry. During her first winter at Annapolis, on March 7, 1862, Sallie gave birth to nine puppies. She became pregnant four more times during her military service. Most of her puppies survived and were sent home to live with the families of the soldiers. In April the regiment went south to fight the rebels for the next three years, and Sallie went everywhere with the men, never deserting them. She fought in all of her regiment's engagements including Cedar Mountain, Second Bull Run, Antietam, Fredericksburg, Chancellorsville, Gettysburg, the Wilderness and Petersburg. Sallie really demonstrated her faithfulness when she got separated from the unit on July 1, 1863, at the Battle of Gettysburg. The soldiers feared that she had been killed, but she was found days later at Oak Ridge, standing guard over the dead and wounded. On May 8, 1864, Sallie was shot in the neck with a bullet that could not be removed but was not fatal. Several months later the wound formed into a cyst and when it burst the bullet was released, leaving a battle scar. On the morning of February 6, 1865, the regiment marched into a fierce conflict near Petersburg, Virginia, and among the dead and wounded lay Sallie, who had been shot in the head. She died that day at Hatcher's Run, Virginia. During her time in the war, she had been honored with a special salute of President Lincoln's hat. A bronze and granite monument of Sallie lying down is at the Gettysburg National Military Park in south-central Pennsylvania.

Salty *see* **Apollo**

A bronze and granite monument to Sallie at the Gettysburg National Military Park in honor of her service during the Civil War (CC BY-SA 3.0).

Sam (Monkey)

Sam's name is an acronym for "School of Aviation Medicine," and his official name was "11X." Sam was an Indian rhesus monkey born in 1957 at the Balcones Research Center. He became part of a secret behavioral and biological research program started at the University of Texas in 1951. Sam was chosen from 15 candidates as a test subject to study what impacts space flight—such as weightlessness, radiation and other conditions—might have on the human body, and also to test the escape system in the launch craft. On December 4, 1959, Sam was launched

Sam, shown here in a photograph dated December 4, 1959, was part of a secret research project at the University of Texas in 1951 to study the impact of space flight on the human body (NASA).

from Wallops Island, Virginia, inside a *Mercury* spacecraft. After one minute traveling at 3,685 miles per hour, Sam's cylindrical capsule separated from the *Mercury* and went on flying for 12 minutes at an altitude of 55 miles before landing safely in the Atlantic Ocean. In late 1959 and early 1960, Sam made flights with a female rhesus monkey, Miss Sam. The two monkeys wore specially wired space suits so that researchers could monitor effects of suborbital flight on their hearts and central nervous systems. A 2017 story by Richard A. Marini in the *San Antonio Express-News* reported that Sam had endured 11 years of further medical research at the School of Aerospace Medicine at Brooks Air Force Base in San Antonio. Sam retired and lived at the San Antonio Zoo until his death on September 19, 1978, at age 21. Apparently after his arrival, he had quite the appetite for oranges. After his death, a necropsy was done and did not find abnormalities due to space flight.

Sandy (Dog)

Sandy, a four-year-old German shepherd, was a messenger dog for the Marines during World War II and was handled by Sgt. Guy C. Sheldon and Sgt. Menzo J. Brown. The unit was sent to the island of New Britain, east of New Guinea, to seize two Japanese landing fields. Making the mission difficult were the pillboxes (concrete dug-in guard posts with holes to fire weapons through) installed along the landing strips. A tropical storm compounded the difficulty; the soldiers' walkie-talkies connecting them to the main unit were not working. This is where Sandy came into the picture. Sergeant Brown placed a message in Sandy's pouch indicating the position of the pillboxes and requesting that the artillery take action. Sandy managed to deliver the message through difficult terrain; thus the pillboxes were destroyed and the Marines

were able to move forward with their mission.

Sarbi (Dog)

Sarbi, a black Labrador retriever and Newfoundland cross, was adopted by the Australian Defence Force to serve as an Explosive Detection Dog (EDD) during the war in Afghanistan. On September 2, 2008, during a Taliban ambush, a gunshot severed the metal clip on her leash, and, being wounded and frightened, Sarbi limped away from the battle; after three weeks she was declared missing in action. Thirteen months later, in November 2009, a healthy Sarbi turned up in an Afghan village, and she was flown back to her Australian base in Tarin Kowt. Australian Prime Minister Kevin Rudd and the International Security Assistance Force head, General Stanly A. McChrystal, got to meet Sarbi on their surprise visit to the Tarin Kowt base. Sarbi was awarded the Purple Cross on April 5, 2011, at the Australian War Memorial. Sarbi died from a brain tumor on March 27, 2015, and her remains were put on display in October 2016 at the Australian War Memorial.

Prime Minister Rudd said of Sarbi's return:

> Things like that, they may seem quite small, but in fact the symbolism is quite strong, and the symbolism of it is us out there doing a job.... We haven't awarded any Australian Victoria Cross for 40 years. Trooper Donaldson stands out there as an Australian hero, and now his dog Sarbi is back home in one piece and a genuinely nice pooch as well.

Satan (Dog)

Satan was a messenger dog, which the French called either "estafattes" or "liaison" dogs. Satan was a greyhound and herding-champion collie mix. During World War I, messenger dogs were often used to send vital messages when other forms of communication were unavailable. Dogs were chosen for the job because they were smaller targets, could run faster than men, travel long distances and detect obstacles such as barbed wire along the way. In the 1916 Battle of Verdun, a French unit was cut off by a German barrage. A dog, a homing pigeon, and seven men were killed attempting to deliver messages about their situation. Losing hope, the French soldiers spotted Satan running towards them wearing a gas mask. Satan's handler, Duval, recognized his dog and as he was urging him on, Satan was shot twice. However having heard Duval's voice, Satan managed to deliver his message to the men in the trenches, which stated, "For God's sake, hold on. We will send troops to relieve you tomorrow." Satan was also carrying two pigeons with him, and the commandant used the pigeons to send two identical messages giving the location of the Germans. One pigeon delivered the message successfully and Verdun was saved. American war reporter Albert Peyson Terhune wrote, "The garrison was able to hold out until reinforcements came all because one hairy mongrel refused to die while his errand was still uncompleted and because he was too loyal to quit." What became of Satan? Some say he died shortly after delivering his message, and other accounts say that he recovered and retired from the military.

Sawbuck (Dog)

John Casler, a private in Company A of the 33rd Virginia Infantry, related the tale of a four-legged Confederate. The Stafford's Guard of the Louisiana Brigade had a very intelligent medium-sized black and white spotted bird dog named Sawbuck. There is no record of Sawbuck's master, but he was probably a local man who had volunteered. During battle, Sawbuck was known to race up and down the front lines

barking at the enemy. He was at Gettysburg cheering on the soldiers. At some point, Sawbuck was shot in the right leg. Recovering from his wound, he was encouraged to remain behind the front lines. Sawbuck recognized every man in his brigade, and, if separated from his unit after a battle ended, he would wait at the side of the road. When he saw one of his men he rushed and kissed him, happy that he was no longer alone. Sawback survived the war.

Scarlett (Cat)

In 1996, an abandoned building in Brooklyn was ablaze when firefighter David Gianelli saw Scarlett, a calico cat, carry her four-week-old kittens, one by one, out of harm even though her eyes were blistered shut and her paws and ears were burned. After pulling them to safety, she lined them up and touched each one with her nose to make sure they were all alive; she then collapsed. The firefighter took the cat and her kittens to an animal shelter, where they were given medical treatment. After the story was featured in worldwide news coverage, the shelter received thousands of requests to adopt Scarlett and her kittens. Scarlett needed continuous care as a result of her injuries. Scarlett was adopted by Karen Weller, who loved and cared for her until the cat passed away in 2008 at approximately 13 years old. The North Shore Animal League created the Scarlett Award for Animal Heroism in Scarlett's honor. The award is presented to animals or humans who have engaged in heroic acts.

Screech (Dog)

Screech became the mascot of the HMCS *Charlottetown* after being found by Seaman John Garland one night in December 1941. He nicknamed the dog after a local Newfoundland drink, and the two became inseparable. On September 11, 1942, the ship was attacked by a German

U-boat and sank. Unfortunately, Garland died trying to rescue Screech, who did survive. The survivors of the attack gave Screech to Garland's mother.

Seabiscuit (Horse)

Seabiscuit was foaled on May 23, 1933, in Lexington, Kentucky, from mare Swing On, and sire Hard Tack, son of Man o' War. He grew up on Claiborne Farm in Paris, Kentucky, which was owned by the Wheatley Stable, and was trained by "Sunny Jim" Fitzsimmons, who also trained Gallant Fox and took him to the United States Triple Crown of Thoroughbred Racing. Fitzsimmons felt that Seabiscuit was too lazy and therefore devoted most of his attention to Omaha, who won the 1935 Triple Crown. Not living up to his racing potential, Seabiscuit's owners sold him for $8,000 to automobile entrepreneur Charles S. Howard, who assigned him to trainer Tom Smith. The years 1936–37 marked the beginning of Seabiscuit's success. Smith assigned Canadian jockey Red Pollard to ride Seabiscuit, and on August 22, 1936, he raced for the first time. During the remainder of 1936, Seabiscuit and Pollard had numerous victories: Detroit Governors Handicap (worth $5,600), Scarsdale Handicap (worth $7,300), Bay Bridge Handicap (won by five lengths), and World's Fair Handicap (Bay Meadows' most prestigious stakes race; Seabiscuit led throughout the race). In 1937, winning 11 of his 15 races, Seabiscuit was becoming the year's leading money winner in the United States. On February 19, 1938, Pollard had a terrible accident and George Woolf became Seabiscuit's new jockey. With his new jockey Seabiscuit lost his first race, the Santa Anita Handicap, referred to as "The Hundred Grander." On November 1, 1938, the "Match of the Century" occurred when Seabiscuit met his relative War Admiral that in 1937 had won the Triple Crown and was given the American Horse of the Year Award at the Pimlico Race Course. There were an estimated 40,000 fans at the track and another 40,000 fans listening on the radio. War Admiral was favored 1–4 with bookmakers, and also favored by writers and tipsters. War Admiral was a fast starter, while Seabiscuit held back and pulled ahead with a late acceleration. From the starting bell, the two horses alternated leads. Two hundred yards from the wire, Seabiscuit pulled ahead and won the race by four lengths. Seabiscuit was named American Horse of the Year for 1938. During a race, Seabiscuit was badly injured with a ruptured suspensory ligament in the front leg, and it was unknown if he would ever race again. Seabiscuit and the injured jockey Pollard recovered together at Howard's ranch. Over the fall and winter of 1939 Seabiscuit improved and was ready to return to horse racing. Seabiscuit and Pollard returned to racing on February 9, 1940, at the La Jolla Handicap at Santa Anita, where Seabiscuit came in third. On his third comeback race he was back to winning, pulling ahead of the field to win the San Antonio Handicap. Seabiscuit came in third in the last race of the season, the Santa Anita Handicap, with a $121,000 prize. On April 10, 1940, Seabiscuit retired to the Ridgewood Ranch in Mendocino County, California. He was horse racing's all-time leading money winner and sired 108 foals. On May 17, 1947, Seabiscuit died, most likely from a heart attack, and he is buried at Ridgewood Ranch. Throughout the years, Seabiscuit has been the subject of numerous books and films including *Seabiscuit: The Lost Documentary* (1939); a Shirley Temple film, *The Story of Seabiscuit* (1949); a book, *Seabiscuit: An American Legend* by Laura Hillenbrand (2001); and a film adaptation of Hillenbrand's book, *Seabiscuit* (2003), which was nominated for the Academy Award for Best Picture.

Seaman (Dog)

Captain Meriwether Lewis and his friend Lieutenant William Clark were chosen by President Thomas Jefferson in 1803 to go on an expedition to find the Northwest Passage, a water route across North America to the Pacific Ocean. The expedition was named the Corps of Discovery by President Jefferson. Captain Lewis bought Seaman (called the "Dog of Discovery"), a Newfoundland, to accompany him on this famous journey. Newfoundlands are intelligent, faithful, good workers and avid swimmers. Both George Washington and Benjamin Franklin owned Newfoundlands. Lewis and Clark and their crew set off on their expedition from Pittsburgh on August 30, 1803 (the official launch was in the summer of 1804, just outside of St. Louis.) During the expedition, Seaman helped the men by retrieving game and guarding their camps against animals, mainly grizzly bears and buffalo. Seaman was also credited with saving lives when a bull charged the camp one night and he was able to steer it away. Fascinated by Seaman, one Native American offered to trade three beaver skins for the dog but was turned down. During the expedition Seaman suffered from heat, mosquitoes, and barbed seeds from the foxtail. He was even bit by a beaver on his back leg, severing an artery. With extraordinary medical care he was able to make a full recovery. In November 1805 Lewis and Clark's expedition reached the Pacific Ocean, and Seaman became the first American dog to cross the continent. Captain Lewis named a Montana creek 'Seaman's Creek' in honor of Seaman, and today it is known as Monture Creek. There is no mention of the dog in Captain Lewis' last entry in his journal on July 15, 1806, so it is not known what exactly happened to Seaman. In his book *A Collection of American Epitaphs and Inscriptions with Occasional Notes* published in 1814, Timothy Alden believes that after surviving the trip Seaman stayed with Lewis. In the book Alden mentions an inscription on a dog collar reading, "The greatest traveller of my species. My name is

SEAMAN, the dog of captain Meriwether Lewis, whom I accompanied to the Pacific Ocean through the interior of the continent of North America." The collar was donated to a museum in Alexandria, Virginia, but was lost in a fire at the museum in 1871. There are several statues and monuments dedicated to Seaman.

Also noted in Alden's book:

> The fidelity and attachment of this animal were remarkable. After the melancholy exit of Gov. Lewis (who died in 1809), his dog would not depart for a moment from his lifeless remains; and when they were deposited in the earth no gentle means could draw him from the spot of interment. He refused to take every kind of food, which was offered him, and actually pined away and died with grief upon his master's grave.

In 2008, Seaman was made the official mascot of Lewis & Clark College's Pioneers. Monument and statue locations of Seaman include: In front of the Custom House in Cairo, Illinois; St. Louis, Missouri; St. Charles, Missouri; Jefferson City, Missouri; Lincoln, Nebraska; the Lewis & Clark Interpretive Center in Sioux City, Iowa; Washburn, North Dakota; Overlook Park in Great Falls, Montana; Fort Clatsop National Memorial in Seaside, Oregon; Cascade Locks Marine Park in Oregon; Columbia View Park in Saint Helens, Oregon; and Sacajawea Center in Salmon, Idaho.

Seaman in Fiction and Nonfiction:

Tall Tails: Cross-Country with Lewis and Clark by Dona Smith (Seaman's "journal" narrates the expedition.)
New Found Land by Allan Wolf (Seaman is the main character.)
The Captain's Dog by Roland Smith (Seaman "narrates" his account of the Lewis and Clark Expedition.)
Seaman: The Dog Who Explored the West with Lewis and Clark by Gail Langer Karwoski
Lewis & Clark: The Journey of the Corps

of Discovery: An Illustrated History by Dayton Duncan and Ken Burns. New York: Alfred Knopf, 1997

"Seaman's Fate" by James Holmberg, Discover Lewis & Clark website (lewis-clark.org)

Secretariat (Horse)

On March 30, 1970, at 12:10 a.m., Somethingroyal foaled a bright-red chestnut colt with three white socks and a star with a narrow stripe; thus was born Secretariat in Caroline County, Virginia. When fully grown, Secretariat was a massive, powerful horse standing at 16.2 hands (66 inches). His chest was so large that he had a custom-made girth, and he was known for his powerful, well-muscled hindquarters. An Australian trainer said, "He is incredible, an absolutely perfect horse. I never saw anything like him." In 1973, he became the first Triple Crown winner in 25 years, setting speed records in all three races. His win at the Belmont Stakes is regarded as one of the greatest races of all time. To sum up his racing career: he won five Eclipse Awards; he received Horse of the Year honors at ages two and three; in 1974 he was elected to the National Museum of Racing and Hall of Fame; and he is on the list of the Top 100 U.S. Racehorses of the 20th Century. Secretariat was sold for a record-breaking $6.08 million at the start of his third year on the condition that he retire from racing at the end of the year. Secretariat's career earnings were $1,316,808. Secretariat was euthanized on October 4, 1989, after being afflicted with laminitis, a debilitating hoof condition. He was buried whole at Claiborne Farm (it is an honor for a racehorse to be buried whole; usually only the head, heart and hooves are buried). Secretariat holds many honors and recognitions and is still recognized as one of the greatest horses in American racing history.

Sefton (Horse)

Sefton was born in 1963 and was euthanized in 1993 at the age of 30. During his life, Sefton won prizes, medals and honors. Some of his prizes were in show jumping and point-to-point races. He was one of the horses used in Buckingham Palace's Changing of the Guard. He was the first British Horse of the Year. There is a wing in the Royal Veterinary College named after him. He was also in active military service for seventeen years, during which time he sustained serious injuries. Traveling to Buckingham Palace, Sefton's unit was hit by an IRA nail bomb (packed with nails, razors and other sharp objects). Sefton was hit in the face and sustained thirty-four wounds. He survived a long, difficult surgery and returned to active service.

Sergeant Bill (Goat)

Sergeant Bill was a Canadian legend owned by a young woman in Broadview, Saskatchewan. His service with the 5th Infantry Battalion of the Canadian Expeditionary Force began in 1914. A train carrying Canadian soldiers on their way to fight in World War I saw Bill and asked a young woman, Daisey, if she would give Bill to them and she agreed. Although intended to be their mascot, Sergeant Bill proved to be very heroic; soldiers who had smuggled Bill to the front lines in Europe told several interesting stories. In one battle, Bill saved the lives of three soldiers by pushing them into a trench just before a shell exploded. Besides his battle wounds, Bill also suffered from shell shock and trench foot during the war. Decorated with three medals, he also proudly wore a fancy blue coat with his sergeant stripes. After the war he was reunited in Winnipeg with his owner, who did not expect to see him again; there he happily lived out his life. After his death, his body was mounted in the Saskatchewan Legislative Building. Eventually his body was returned to his home and is on display in the Broadview Historical Museum. Bill's story inspired a romantic comedy short film, which was screened at Yorkton Film Festival. Children's book author Mireille Messier wrote a book about Bill, and his story was featured on the Travel Channel's *Mysteries at the Museum*.

Sergeant Reckless (Horse)

Sergeant Reckless (AKA "America's War Horse") was recognized in *Life* magazine's special collector's 1997 edition, "Celebrating our Heroes." He was a small, red Mongolian mare who became known as the greatest war horse in American history. Born with the name Ah-Chim-Hai ("flame of the morning" in Korean), Sergeant Reckless started out as an award-winning racehorse. After North Korea invaded the south in 1950, Ah-Chim-Hai and her owners fled to the Pusan perimeter, where they made a meager living using her as a workhorse until their return to Seoul in 1952. Due to the hardships of war, Sergeant Reckless' family was forced to sell her. Second Lt. Eric Pedersen was looking for a pack animal to haul ammunition up the Korean mountains. Pederson saw Ah-Chim-Hai's potential, bought her and renamed her Reckless. Reckless began her service with the Marines of the Recoilless Rifle Platoon, 5th Marine Regiment, in October 1952. She was trained to follow trails, carry a packsack, and stay calm in dangerous situations. Reckless proved her worth on the battlefield. During the Battle for Outpost Vegas in 1953, despite being wounded, Reckless made fifty-one trips in one day,

Sergeant Reckless, shown here in retirement ca. 1954–1955 at Camp Pendleton, California, earned many awards honoring her invaluable service with the Marines of the Recoilless Rifle Platoon, 5th Marine Regiment (photograph by Andrew Greer, United States Marine Corps).

covering thirty-five miles over hilly terrain while carrying nine thousand pounds of ammunition to the combat zone and bringing wounded soldiers back for medical care. Upon her retirement, Reckless earned many awards including: promotion to staff sergeant; two Purple Hearts; the Marine Corps Good Conduct Medal; the National Defence Service Medal; the Korean Service Medal; and unit citations. In 2016 she was awarded the Dickin Medal posthumously for her service in the Battle for Outpost Vegas. Reckless' service is commemorated by a memorial statue at the National Museum of the Marine Corps in Triangle, Virginia.

Sergeant Stubby (Dog)

Sergeant Stubby was the first and most famous dog used by the United States in World War I, serving for 18 months in France with the 102nd Infantry. Also this pit bull (the breed today is given such a negative connotation) was the only dog promoted to sergeant during combat. In 1918, Stubby made front-page news in the United States. His military career began in 1916 or 1917, when he was found by some soldiers training at Yale Field in New Haven, Connecticut. Stubby's military tasks included detecting sounds and smells from a distance in order to warn the soldiers of artillery attacks and mustard gas, and saving the lives of many soldiers by locating and comforting the wounded. He can take the credit for capturing a German spy in the Argonne. During his service, he took part in four offensives and 17 battles and survived several wounds. At the end of the war, Stubby became a celebrity in the United States, marching in parades

Serving for 18 months in France with the 102nd Infantry, Sergeant Stubby (shown here between 1918 and 1921) became the most famous dog used by the United States in World War I (Connecticut State Military Department).

and meeting U.S. presidents. He became the mascot of the Georgetown Hoyas and was made a life member of the American Legion, the Red Cross and the YMCA. In his memory, his stuffed body is displayed at The Price of Freedom: Americans at War exhibit at the Smithsonian National Museum of American History.

Sevastopol Tom *see* **Crimean Tom**

Shamu (Orca)

In the 1960s in San Diego, SeaWorld's main attraction was an orca named Shamu (the first orca to survive more than thirteen months in captivity). After her death in 1971, the name Shamu continued to be used for other killer whales in different SeaWorld parks.

Sheila (Dog)

For her bravery, Sheila was the only civilian awarded the PDSA Dickin Medal, which is the animal equivalent of the Victoria Cross. On December 16, 1944, around lunchtime in Dunsdale, England, John Dagg and his collie Sheila were tending to the sheep when they heard a plane crash on the top of the hill. The plane, a Boeing B-17 Flying Fortress from the U.S. 8th Air Force, was carrying a payload of bombs. Through blowing snow Sheila led Dagg and another shepherd, Frank Moscrop, who was nearby, up the hill to the aircraft. They helped four surviving airmen down the hill, with one being badly injured; two airmen died and three others, believing they were the only survivors, made their own way down the hill. The bombs exploded shortly after Dagg's group made it to his cottage. If not for Dagg, Sheila, and Moscrop, Lt. George Kyle, Sgt. Howard Delaney, Sgt. George Smith and Sgt. Joel Berly may have perished that day. Dagg and Mascrop were awarded the British Empire Medal. In 1946, Tibbie, one of Sheila's puppies, was sent to the family of Sgt. Turner, one of the fatalities of the B-17 crash.

Sheila (Elephant)

A Northern Ireland woman, Denise Weston Austin ("the Elephant Angel"), rescued baby elephant Sheila from the Belfast Zoo by keeping the elephant in her backyard during World War II. Her decision came after officers from the Royal Ulster Constabulary shot dead 23 animals at the zoo. The Constabulary had been acting on orders from the British Ministry of Public Security, which feared that the animals would escape during the bombing and wreak havoc. Ms. Austin, one of the first female keepers at the zoo, saved young Sheila by sneaking her out every night and walking her home, and then returning her to the zoo every morning. Sheila was given hay from the family farm. No one knew about Sheila's second home until she chased after a dog into a neighbor's garden, breaking the fence. Sheila then remained at the zoo full time. Denise would visit Sheila at night, rubbing her ears to keep her calm. Sheila lived another 25 years until her death at the zoo in 1966.

Shenkin (Goat)

Shenkin died at the age of 13 after holding the record (12 years) as the longest-serving regimental goat. He was chosen from the queen's herd of Royal Windsor Whites. (Taffy, presented to the Royal Welsh by Queen Victoria after the Crimean War, had been the first in a long line of army goats.) Shenkin was a very busy goat, involved with charity work, weddings, and school visits; he led the Wales rugby team onto the field at Millennium Stadium. He also met the queen, appeared on *Blue Peter* (a British children's television program), and was invited inside 10 Downing Street.

Shep (Dog)

Shep, "the Denver-Boulder Turnpike Dog," earned his place in history when the Colorado Department of Transportation named him the unofficial mascot of the Denver-Boulder Turnpike. Shep's story begins in 1950 when, as a stray puppy, he wandered onto the construction site of the turnpike and the workers began to feed him. The turnpike opened in 1952 and Shep began to beg for food from the tollbooth attendants. On one cold night Shep found his way, via an attendant, to a warm place inside the tollbooth, and soon thereafter he was friends with all the attendants. He also became loved by all the motorists, who often paid a little extra toll to contribute to the purchase of dog food, toys, and treats. Some also stopped to have their picture taken with him. Sometimes Shep would go exploring, and one day in 1958 he returned with a mysterious gunshot wound. Shep was treated for free by veterinarian Clyde Brunner, who also donated his services for the rest of Shep's life. On August 3, 1964, Shep was put to sleep after beginning to lose his sight and having difficulty getting around. Shep was buried next to the U.S. 36 on-ramp with two headstones, "Shep 1950–1964, Part Shepherd–Mostly Affection," and "Our Pal." In 2009, Shep's grave was moved to Zang Spur Park next to the Bloomfield Depot Museum. Shep was truly loved. Jane Spain, who helped lead the effort to move Shep's grave, said, "He always wagged his tail at everyone who came by, and he was always happy. For everyone that went back and forth on the turnpike, he was their favorite thing."

Shep (Dog)

The name Shep was given to a herding dog who appeared one day in 1936 at the Great Northern Railway in Fort Benton, Montana. Shep appeared at the station to watch a casket being loaded onto the train bound for the eastern United States. After that, Shep kept returning to the station to wait for every incoming train. After some time, the station employees realized that it had been the body of Shep's master in that casket, and Shep was waiting for him to get off the train. Shep kept up his vigil for almost six years, until he was hit by a train on January 12, 1942. Nearly everyone in Fort Benton attended his funeral, and his grave is on a hillside overlooking the town. Shep's story was written in *Ripley's Believe It or Not*. His story is told in a historical fiction novel, *Shep: Forever Faithful* by Stewart H. Beveridge and Lee Nelson. In 1994, a bronze sculpture of Shep by Bob Scriver was unveiled at Fort Benton.

Shrek (Sheep)

Shrek was named after the popular ogre of the movie *Shrek*. Shrek was a male Merino sheep born in 1994 in Bendigo Station near Tarras, New Zealand. In 1998, he went to the high country of Central Otago on the South Island and spent the next six years living off the land. In 2004, Shrek was discovered in a rock cave in the Bendigo high country by some workers out looking for sheep. He had not been sheared for six years—grooming is normally done annually. Over 75 percent of his body was covered in wool, and his fleece weighed 60 pounds, whereas an average weight for a Merino fleece is 9.9 pounds. Shrek became a national icon. He met the New Zealand prime minister and was sheared on national television. His fleece was auctioned off for children's medical charities, raising $150,000. Shrek was euthanized on June 6, 2011, on the recommendation of a veterinary surgeon. His owner, John Perriam, said, "He had an unbelievable personality. He loved children and was really good with the elderly in retirement homes."

Siam (Cat)

In 1879, President Rutherford B. Hayes received the first documented Siamese cat to reach the United States. A cat named Siam had been sent by the American Counsel in Bangkok. The relatively brief life and activities of Siam were well-known during her time as "First Cat." Siam was elegant and slender, with long legs and bright blue eyes. Lucy Hayes first named her Miss Pussy, but soon changed her name to Siam. The cat was a favorite of the president's daughter, Fanny. Unfortunately, Siam became sick several months after arriving. The president's own physician examined her, but Siam did not recover.

Silver (Dog)

Silver was killed in action by an enemy grenade on February 17, 1945. However, she is credited with preventing casualties by alerting soldiers prior to a bayonet attack.

Simon (Cat)

A cat as a war hero? Simon (official title: Able Seacat Simon) was born in 1947 in Hong Kong and became a famous and honored cat. Found abandoned, Simon was adopted by Seaman George Hickinbottom and brought back to his British ship, the HMS *Amethyst*. Typically, cats were valued for catching rats on board the ships; however Simon went above and beyond this task. The *Amethyst's* mission was to protect the British people living on the Yangtze River in China. While in China, the ship was attacked, some seaman were killed and the ship sailed to shore. The men who tried to escape were killed instantly, leaving about 50 men on board, some injured; Simon was also injured. For the next three months, the ship remained heavily guarded by the enemy. Simon remained on board and caught at least one rat a day, which helped to preserve the sailors' food supply; he was also a great comfort to the injured men. Finally, after 101 days they were able to escape and sailed back to Hong Kong. Simon became famous after the newspapers and radio reporters caught wind of the story. Simon is the only cat to be awarded the Dickin Medal. Simon died on November 28, 1949, a month before he was to be presented with the medal. The crew of the *Amethyst* buried Simon with full naval honors in a pet cemetery near London, England. Simon's photograph and story appeared in *Time* magazine. Simon also was awarded a Blue Cross medal, the *Amethyst* campaign medal, and the rank of "cat officer." Other honors include: a bush is planted in Simon's honor at the National Memorial Arboretum in Staffordshire, and writer Paul Gallico dedicated his novel *Jennie* to Simon.

Simon's gravestone reads:

> IN
> MEMORY OF
> "SIMON"
> SERVED IN
> H.M.S. AMETHYST
> MAY 1948–NOVEMBER 1949
> AWARDED DICKIN MEDAL
> AUGUST 1949
> DIED 28TH NOVEMBER 1949
> THROUGHOUT THE YANGTZE INCIDENT
> HIS BEHAVIOUR WAS OF THE HIGHEST
> ORDER

Sinbad (Dog)

Sinbad is famous in history for being a mascot of the U.S. Coast Guard during World War II; he liked to have fun and keep the crew on the USS *Campbell* smiling. In 1937 Sinbad was intended to be a gift to a girlfriend of one of the sailors of the USS *Campbell*. However she was not able to keep the dog, so that night the sailor smuggled Sinbad aboard the ship. However, his barking the next morning made his presence known to all. Sinbad quickly became an official crew member—he provided a paw print signature on his enlistment papers; had his own service record, Red Cross and service IDs; and had a bunk. Sinbad was the ship's mascot for over 11 years, during both peace and war time. The ship was awarded in 1943 for battling six enemy submarines during 12 hours and sinking the German submarine U-606. The crew believed that as long as Sinbad was aboard, the ship was unsinkable. Viewed as a capable sailor, Sinbad won six campaign ribbons and five battle stars, but while on liberty he let loose with the crew, accompanying them to bars and nightclubs. The captain promoted Sinbad to K9C (Chief Petty Officer, Dog); however, a few weeks later, he lost his promotion for insubordination. Eddie Lloyd, editor of *Coast Guard* magazine, described Sinbad as "a salty sailor but he's not a good

Sinbad, mascot of the U.S. Coast Guard during World War II, and World War I dog hero Sergeant Stubby, were the only two animals classified as non-commissioned officers by the U.S. military.

sailor. He'll never rate gold hash marks nor good conduct medals. He's been on report several times and he's raised hell in a number of ports. On a few occasions, he has embarrassed the United States Government by creating disturbances in foreign zones. Perhaps that's why Coast Guardsmen love Sinbad, he's as bad as the worst and as good as the best of us." Honorably discharged in 1948, Sinbad lived out his remaining years at Barnegat Coast Guard Station in New Jersey and passed away on December 30, 1951. He was buried beneath a granite monument at the base of the station's flagstaff. Sinbad and Sergeant Stubby (World War I dog hero) were the only two animals classified as non-commissioned officers by the U.S. military. (Afterwards there were regulations prohibiting such classifications.) There is a statue of Sinbad on the mess deck of the successor to the *Campbell*, "famous-class" medium-endurance cutter USCGC *Campbell* (WMEC-909). *The Boston Globe's* Martin Sheridan described Sinbad in a December 1943 *Life* magazine article as "liberty-rum-chow-hound, with a bit of bulldog, Doberman pinscher, and what-not. Mostly what-not." Sinbad became more celebrated with the publication of George F. Foley's *Sinbad of the Coast Guard*; Sinbad traveled with the book signing tour.

Sir Briggs (Horse)

The gravesite of Sir Briggs can be found in the Cedar Garden at the Tredegar House in Newport, Wales. There is a monument at his grave, which reads: "In memory of Sir Briggs. A favourite charger: He carried his master, The Honourable Godfrey Morgan Captain 17th Lancers boldly and well at the Battle of the Alma, in the first line in the Light Cavalry Charge of Balaclava and at the Battle of Inkerman 1854. He died at Tredegar Park 9th February 1874 aged 28 years." For more than

500 years, the Morgan family (later known as Tredegar) owned 40,000 acres. Captain Godfrey Charles Morgan and his horse, Sir Briggs, were involved in heavy fighting in the Crimean War. In 1853, Sir Briggs was transported to Russia from Wales to fight at the Light Cavalry Charge of Balaclava and at the Battle of Inkerman. Sir Briggs was the war horse that inspired Tennyson's "Charge of the Light Brigade."

Sirius (Dog)

Four-year-old golden Labrador retriever Sirius is the only known canine casualty of the September 11, 2001, tragedy. Sirius, a police dog, worked with his handler, David Lim, for the New York/New Jersey Port Authority. Their job was to inspect vehicles coming into the World Trade Center. That morning the two were in the South Tower basement when the first plane hit the North Tower. Lim put Sirius in his crate and promised he would be back. For five hours, Lim was trapped in debris in the North Tower, and the South Tower collapsed on Sirius. Workers found Sirius' remains on January 22, 2002. Lim had his remains cremated and kept the ashes. The War Dog Memorial in the Hartsdale Cemetery and Crematory in New York installed a marker for Sirius. Sirius also has a dog run in his honor in Battery Park City's Kowsky Plaza.

Siwash (Duck)

Siwash, a New Zealand duck, was the mascot of the First Battalion of the Tenth Marine Regiment and later the entire Second Marine Division during World War II at Tarawa, Saipan, and Tinian. Supposedly, a Marine won the duck in a poker game in New Zealand. Siwash went ashore with the Marines at the Battle of Tarawa in 1943. There she was cited by fellow Marines for fighting a Japanese-owned rooster. Siwash was skilled at the sport of drinking beer.

After the war, she lived on a farm before becoming a Marine recruiter during the Korean War. She died in 1954 at the Lincoln Park Zoo in Chicago.

Skippy (Dog)

Skippy was a wire fox terrier owned and trained by Henry and Gale East, who were assisted by the Weatherwax brothers and Frank Inn. Skippy began his training at just three months old and was considered to be one of the most intelligent animal actors at the time; he was able to respond to verbal commands and hand cues and had an acting range that included acting afraid, mad or impersonating a bloodhound. Skippy is most known for his role as Asta, the pet of Nick and Nora Charles in the first three *Thin Man* movies. Other terriers performed the role of Asta in the sequels, *After the Thin Man, Another Thin Man, Shadow of the Thin Man, The Thin Man Goes Home, Song of the Thin Man,* and in the 1950s television show *The Thin Man.*

In his roles Skippy had inconsistent billing, as his filmography demonstrates, being credited as Skippy, Asta, or not at all.

The Half-Naked Truth (1932), uncredited
Fog Over Frisco role: Ragsy (1934)
The Thin Man role: Asta (1934), credited as Asta
The Big Broadcast of 1936 (1935), uncredited
Lottery Lover role: Pom Pom (1935), credited as Skippy
It's a Small World (1935)
The Daring Young Man (1935)
After the Thin Man role: Asta (1936), credited as Asta
China Passage (1937), uncredited
Sea Racketeers role: Skipper (1937), credited as Skippy
The Awful Truth role: Mr. Smith (1937), uncredited

Bringing Up Baby role: George (1938), credited as Asta

I Am the Law role: Habeas (1938), credited as Asta

Keep Smiling role: Mr. Skip (1938), credited as Skippy

Topper Takes a Trip role: Mr. Atlas (1939), credited as Skippy

Another Thin Man role: Asta (1939), credited as Asta

I'm Still Alive (1940), uncredited

Shadow of the Thin Man role: Asta (1941), credited as Asta

The Thin Man Goes Home role: Asta (1944), Skippy's successor

Song of the Thin Man role: Asta (1947), Skippy's successor credited as Asta

Sky Queen *see* **Elm Farm Ollie**

Sleepy Tom (Horse)

Sleepy Tom was born in 1866, and despite having the bloodline of the two American Thoroughbreds Eclipse and Messenger, he had none of their spirit or speed and in fact, his drowsy way of moving gave him his name. Sleepy Tom was purchased by Stephen C. Phillips in 1878 for $7.50 and a jug of whisky. When Phillips sold him, he was unaware that he was selling him into brutal slavery. Years later, Phillips purchased the mistreated horse and nursed him back to health and strength. Phillips discovered that Sleepy Tom had a great talent as a pacer for harness racing. When Sleepy Tom made his debut on the track, the crowd did not take him very seriously, and quoting one of his biographers, Sleepy Tom was "the laughing stock of the crowd." However, the laughter stopped as the crowd watched the horse gaining the lead with his strong legs not breaking their stride. What is more miraculous is that Sleepy Tom was running completely blind. He had started to go blind in 1874 at age seven

due to the brutal treatment from previous owners. Winning race after race, he also set world records: the first was a mile in 2 minutes, 22 seconds; when he was thirteen he beat his own record, running a mile in 2 minutes, 12 seconds. The famous Sleepy Tom died at twenty years old in a barn fire in 1886.

Slippers (Cat)

Slippers, a bluish-gray, six-toed tabby cat, belonged to President Theodore Roosevelt. Slippers would often wander off but always seemed to return to the White House whenever a state dinner was happening. At one state dinner in 1906, the procession led by Roosevelt, who was escorting the wife of an important British ambassador, came across a snoozing Slippers sprawled out in their way. Roosevelt and his guest went in an elliptical direction around the cat, as did everyone else.

Smokey the Bear *see* **Hotfoot Teddy**

Smoky (Dog)

Smoky (nicknamed "Yorkie Doodle Dandy") was a tiny Yorkshire terrier (4 pounds and 7 inches tall) born in 1943 and found in 1944 by Ed Downey in a foxhole in the New Guinea jungle—abandoned, dirty, and hungry. Initially the soldiers thought that she belonged to the Japanese; however after taking her to a prisoner-of-war camp they found that she did not understand commands in either English or Japanese. Not fond of dogs, Downey sold her to Corporal William Wynne, a 22-year-old from Ohio, and he named her Smoky. Smoky and Corporal Wynne backpacked through the jungle for the next two years, sharing a tent and rations and surviving air raids, typhoons and twelve combat missions together. Smoky served in the South Pacific with the 5th Air Force, 26th Photographic

Reconnaissance Squadron. Smoky can take credit for participating in 12 air/sea rescue and reconnaissance missions and 12 combat missions (she was awarded eight battle stars); surviving 150 air raids on New Guinea and a typhoon at Okinawa; parachuting out of a tree 30 feet in the air; warning the soldiers of incoming fire with her exceptional hearing and sense for danger; and being the first war-therapy dog. In 1944 she was named the "Champion Mascot in the Southwest Pacific Area" by the magazine *Yank Down Under*. In January 1945, she saved the lives of 250 men in 40 planes by helping to install a communications cable through a 70-foot pipe, eight inches in diameter, under the runway at an airbase at Luzon. Thanks to her small size, this brave dog completed the mission, accomplishing in minutes what would have taken the men eight days. Smoky also saved Wynne's life near the Philippines on a tank landing ship by warning Wynne to duck when fired upon by enemy planes, which hit eight men close by. Wynne deemed Smoky an "angel from a foxhole." When Wynne was hospitalized with dengue fever, he realized how good Smoky was at charming and cheering up other patients. Therefore, Wynne began teaching her tricks, which included walking blindfolded on a tightrope, riding a handmade scooter and spelling her name by picking out the correct letters as they were called out. Smoky became a national celebrity after being featured along with Wynne in a front-page story in *The Cleveland Press* on December 7, 1945. Smoky and Wynne had their own television show in Cleveland called *Castles in the Air*, and she performed on 42 live television shows, never repeating a trick. Smoky entertained troops with Special Services and patients in hospitals by performing her tricks. According to Animal Planet, Smoky is credited as being the first therapy dog on record; she worked for 12 years during and after World War II. By 1947, civilians had donated about 700 dogs to be used as therapy dogs. Smoky died in her sleep at the age of 14 on February 21, 1957. There are six memorials honoring this famous war dog. One such memorial was unveiled on Veterans Day, November 11, 2005. It is a life-size bronze statue depicting Smoky sitting in a GI helmet on a two-ton blue granite base and is placed on top of Smoky's grave. This monument is dedicated to "Smoky, the Yorkie Doodle Dandy, and the Dogs of All Wars."

Snowball (Cat)

The first time non-human DNA was used to solve a murder occurred in 1994 in the case of 32-year-old Shirley Duguay. This murder took place on small Prince Edward Island off the coast of Canada. The Royal Canadian Mounted Police found a body buried in a shallow grave along with a blood-soaked jacket. The blood belonged to the victim, but they also discovered long white cat hairs. Police knew that the victim's common-law husband owned a white cat named Snowball. The authorities obtained Snowball's blood to perform a DNA test. After many calls and hours of research by the police, an institution finally agreed to test the sample. Previously no lab had ever tested a pet's DNA. The police collected blood samples from about 20 cats to make sure that felines did not have similar DNA. The cat hairs did belong to Snowball, which was enough to convict his owner and the victim's husband, Douglas Beamish. Since then pet DNA has been useful in solving many violent crimes.

Snowflake (Gorilla)

After his family was killed, Snowflake was captured in the Rio Muni region of Equatorial Guinea on October 1, 1966, at about age two. Originally named Nfumu Ngui ("white gorilla" in Fang language) by his captor, he was then nicknamed

"Floquet de Neu" (Catalan for "Little Snowflake") by his keeper, primatologist Jordi Sabater Pi. Snowflake was a western lowland gorilla with non-syndrome albinism. It is believed that his parents were uncle and niece. In 1966, Snowflake was sent to live at the Barcelona Zoo. Snowflake fathered twenty-two offspring and none were albino. In 2001, Snowflake was diagnosed with an unusual form of skin cancer, and on November 24, 2003, he was euthanized. Snowflake was the zoo's most famous resident for almost forty years.

Snowman (Horse)

Snowman was a champion jumper during the 1950s, but it had been a long road getting there. Snowman was referred to as "The Cinderella Horse" with his "rags to riches" story. A jumper is a Thoroughbred with long limbs—jumping gracefully to the ground like the famous jumpers Andante, who won Horse of the Year three times; Diamant; and beautiful First Chance, all beaten by Snowman. He started out as a broken-down, big, heavy, overworked and underfed horse headed to becoming dog food. One February day in 1956, Snowman was being loaded onto a truck, and a gleam in his eye caught the attention and sheer pity of Long Islander Harry de Leyer, who bought Snowman for eighty dollars. The horse was so filthy that it took five washings to clean him; he had sores and deep grooves on his skin due to the plow harness he wore. Not being able to use the horse as he had hoped at the Knox School for Girls, where he was riding master, de Leyer

Snowman, pictured jumping here in October 1962 with his owner Harry De Leyer, who took him from a broken-down horse to a champion jumper during the 1950s (photograph by William C. Greene, *New York World-Telegram* and the *Sun* Newspaper Photograph Collection. Library of Congress).

sold Snowman for $140 to a doctor across town. Even though the doctor treated him well, Snowman showed his great loyalty to de Leyer. This big horse jumped over the doctor's fence, returning to de Leyer, who took him back; another bar was added to the fence and once again he returned to de Leyer, who took him back a second time. This pattern went on for quite a while with the fence rising higher and higher, and the horse continued to jump over the fence. Eventually, seeing that Snowman was not going to stop, de Leyer purchased him from the doctor. With all this jumping over the fence, Snowman was a natural-born jumper, de Leyer realized. Within a few years, Snowman had won three prestigious open jumping titles—the Professional Horsemen's Association of America Champion, the American Horse Shows Association Horse of the Year, and the National Open Jumper Championship—and won thousands of dollars. De Leyer was offered $100,000 for the horse, which he refused. It was Snowman's apparent love and loyalty towards the first person who had been kind to him that shot the horse to stardom. In the fall of 1974, at the age of twenty-six, Snowman was humanely euthanized by de Leyer due to kidney failure. In 1992, Snowman was inducted into the United States Show Jumping Hall of Fame. A number one *New York Times* bestseller, *The Eighty Dollar Champion: Snowman, the Horse That Inspired a Nation* by Elizabeth Letts, was published by Random House in 2011. A documentary movie, *Harry & Snowman*, was made in 2015.

Socks (Cat)

This little stray cat with white paws was spotted in the garden of a little girl's piano teacher in Little Rock, Arkansas, in 1991. This little cat became famous as the White House Cat during Bill Clinton's presidency. However, when the Clintons acquired their Labrador retriever, the two pets did not get along and had to be kept separated. After Bill Clinton left office and moved into a smaller house, this arrangement could not continue for lack of room. Therefore, Socks lived out the rest of his life with Clinton's secretary Betty Currie. In February 2009, Socks developed cancer of the jaw and was euthanized at age 18.

Spike (Dog)

Spike was a Mastiff/Labrador retriever mix, rescued as a puppy from a shelter in Van Nuys, California, by animal trainer Frank Weatherwax. Spike is best known for his starring role as Old Yeller in the 1957 Disney film of the same name. Spike also appeared as Patrasche in *A Dog of Flanders* (1959), King in *The She-Creature* (1956), and Pete in *The Silent Call* (1961). Spike also appeared in several television series: *The Mickey Mouse Club*, *Lassie*, and *The Westerner*. Spike died in 1962 at 9–10 years old.

Spot (Dog)

General Charles Townshend's mascot Spot was a fox terrier earning the nickname "Spot o'Kut" because the general was able to hold the city of Kut al-Amara for months before being defeated. The general had led his infantry, the 6th Poona Division, in a campaign near Baghdad and was defeated by the Germans and Turks. Spot was also captured and stayed with Townshend until the general was sent to Constantinople. Turkish law prohibited the general from taking his dog with him, so Spot was sent down the Tigris River and given to the British. He was then sent to England, where he lived with a friend of General Townshend.

Squeak (Dog)

Since gaining independence in 1980, Zimbabwe had difficulty establishing stability, particularly in the early 21st century, with land-redistribution programs igniting violence. On March 18, 2002, this violence came to a farmer named Terry Ford, who was bludgeoned to death on his property. Ford's faithful 14-year-old Jack Russell terrier named Squeak made news when he refused to leave his owner's slain body. Ford's sister in Melbourne, Australia, only learned the fate of her brother when she saw news footage of Squeak guarding his body. Ford's fiancée removed Squeak from his vigil and took him to the funeral.

Star (Horse)

Sybil Ludington was the oldest of Colonel Henry Ludington's twelve children. The colonel had been given the rank in 1776 and was given a regiment in Duchess County. Sybil excelled as a horse rider. At 15, her father gave her a horse, which she named Star due to the white blaze on his forehead. She trained Star and watched her father train the militia at their farm, and assisted by running errands and following troops home. In April 1777, Sybil was 16 years old. A message arrived about upcoming troubles and her father needed to remain at the farm and direct the troops as they arrived. Sybil was the logical choice to ride out and muster the militia. British General William Tyron had landed off the coast, moving inland unopposed. His 2,000 soldiers attacked and set fire to homes in Danbury, Connecticut. Sybil rode about a 40-mile route on rough roads to get out the word. She has been called "the female Paul Revere." There is some doubt about the full truth of this story. However, there is a statue of her and Star and in 1975, the U.S. Postal Service issued a stamp of Sybil.

Stickeen (Dog)

In 1880, naturalist and explorer John Muir set out in a canoe with a crew of Native Americans from Fort Wrangell in southeastern Alaska to explore the coastline. A friend of Muir's from the fort, the Reverend S.H. Young, joined him for the journey, bringing along his small black dog, Stickeen (named after a tribe living near the fort). Muir was not happy about Stickeen accompanying them. He thought the dog would be cold and miserable. Stickeen was a "poor silly thing," a toy dog, and would have to be cared for like a baby. Muir initially described the two-year-old dog as "a queer character—odd, concealed, independent, keeping invincibly quiet, and doing many puzzling things that piqued my curiosity." Stickeen enjoyed the icy waters by leaping out of the canoe first when approaching a landing place. He never obeyed an order and no hardship seemed to faze Stickeen. Muir came to respect the animal's intelligence and ability to fend for himself. Stickeen took a liking to Muir and began following him on all hikes, no matter the terrain. One day Stickeen followed Muir when he set out from one of the expedition's campsites for a day-long hike across a large glacier. Muir left early in the morning on a stormy day and ordered Stickeen to return to the camp, but he came anyway. Muir was worried because he and Stickeen were traveling up and down icy slopes and leaping over deep crevices. Daylight was fading and the crevices were getting wider. Then they came upon a crevice neither could leap over. Muir decided to cross this ice bridge by using his pickaxe to notch out some steps in the ice wall, leading down to the ice bridge. He straddled the ice, inched his way across the crevice and climbed to safety. Muir realized that Stickeen was terrified of going on the ice bridge and finally ordered him to come to him. Stickeen inched down the icy steps, crept down the ice bridge and climbed up the notches in a hurry. The moment Stickeen knew he was safe, he became excited with happiness. There were more crevasses but none as bad and they made it safely back to camp. Muir wrote, "I have known many dogs, and many a story I could tell of their wisdom and devotion; but to none do I owe so much as to Stickeen. At first the least promising and least known of my dog-friends, he suddenly became the best known of them all. Our storm-battle for life brought him to light, and through him as through a window I have ever since been looking with deeper sympathy into all my fellow mortals."

Strelka see Belka and Strelka

Stubby (Dog)

Stubby, a bull terrier, was not a trained war dog but became a hero in his own right during World War I. Not much information is known about him. One day he simply showed up at the Yale field, so it was not known where he came from, how old he was, or to whom he belonged. Corporal Robert Conroy became Stubby's friend (handler) and smuggled him onto the transport to France with the aid of an MP. Stubby was a Marine accompanying Corporal Conroy when his unit was sent to the front lines. Not being trained for war, he saw it as a game of dodging shells and jumping into foxholes. His sensitive ears could hear the shells before the soldiers could, and they began watching him taking cover as a warning that they should do the same. He was not frightened by the shells, but he showed great concern for the wounded soldiers. Wanting to help the men, he would go from one to another, whining and licking their hands and face. Stubby would get help by finding Conroy or the nearest soldiers, and he would bark incessantly, racing back and forth and biting at their pants and jackets until they followed him to the wounded. Once again Stubby saved the lives of Conroy and the other Marines by waking them up and alerting them to a gas attack. It is

not known how one night Stubby sensed a spy nearby and captured him by sinking his teeth into the man's pants. After the war, Stubby went home with Conroy and became a life member of the American Red Cross and the Eddy-Glover Post No. 6 of the American Legion in New Britain, Connecticut. He would lead the American Legion parades wearing his embroidered service jacket made for him by the women of Château-Thierry. He was even welcomed by the proprietor of New York's Hotel Majestic, where dogs were never allowed. Stubby had one more act of bravery several years after returning to civilian life. Conroy and Stubby were taking a nostalgic walk through the streets of Paris. They were preparing to cross the street when suddenly Stubby tore his leash out of Conroy's hand, ran to the opposite curb and pushed a young girl out of the way of a speeding taxi; both the girl and her rescuer were unharmed. Stubby had saved the girl's life. Stubby died on April 4, 1926, and a plaster cast was made of his body to hold his ashes and presented to the Smithsonian Institution in Washington, D.C.

"Stumpy" John Silver (Pigeon)

This homing pigeon was hatched in January 1918 in a dugout just behind the battle lines in France. During the Meuse-Argonne offensive, "Stumpy" was one of the most active pigeons in the Army. His dodging skills were apparent in his many exciting flights delivering messages. On October 21, 1918, John Silver was released from a dugout in the Meuse-Argonne with a message for headquarters 25 miles away. Furious bombing was going on, but through the fire John gained his bearings and made it to headquarters. Men in the trenches witnessed a shell explode near the pigeon, and the blast tossed him upward and then downward. He regained consciousness and appeared to be a terrible sight. A bullet had ripped his breast, bits of shrapnel ripped his body and his right leg was missing, but the message remained intact. Weeks of nursing restored his health but he lost the leg and an eye. The pigeon became a war hero and earned the name John Silver after the pirate in *Treasure Island*. He was retired from active service and in 1921 was assigned as a mascot to the 11th Signal Company, U.S. Army Signal Corps, Hawaii. John died on December 6, 1935, at the age of 17 years and 11 months. Since at least the mid–1930s, many people have called him "Stumpy" John Silver. Some people don't approve and it is a source of contention. His stuffed body is displayed at the National Museum of the U.S. Air Force at Wright-Patterson Air Force Base.

Stymie (Horse)

Stymie, a chestnut horse, was born on April 4, 1941, on King Ranch in Texas. His owner, Max Hirsch, did not see much potential in Stymie for a career as a racehorse. Esteemed trainer Hirsch Jacobs bought Stymie for $1,500 for his wife Ethel on June 2, 1943. At that time, the horse's racing record was seven wins out of fifty starts. At two years old Stymie had lost all but one of his races, but by the age of three he came in second in the Wood Memorial Stakes and third in several more races. In 1945, racing was shut down for four months by the U.S. government, and after seven months Stymie came back winning eight races, placing second in three races and third in five races. At ages five, six, and seven he continued winning until his racing career ended in 1949, when he fractured a bone in his right front leg. To sum up his racing career, out of 131 starts he won 35, placed in 33, and showed in 28 with winnings totaling $918,485. In 1975, Stymie was inducted into the National Museum of Racing and Hall of Fame. Stymie died in 1962. Jacobs and his wife named their breeding operation in Sparks, Maryland, Stymie Manor.

Su-Lin (Panda)

The first live giant panda seen in the western world arrived in New York from China on December 23, 1936, with explorer Ruth Harkness. Originally it was her husband who was searching for a panda, but he died and Ruth flew to China and finished the expedition. In his book *The Last Panda*, biologist George Schaller describes how Harkness' expedition created a "happy furor" across the country. *The New York Times* printed a delightful headline, "Baby Panda Here, Enjoys Its Bottle." Su-Lin was about 9 weeks old, weighed 5 pounds, and was thought to be a female. Su-Lin was named after Su-Lin Young, the sister-in-law of Harkness' Chinese-American expedition partner Quentin Young. They were unaware that the baby panda was really a male. In April 1937, the panda was purchased for just under $9,000 by Chicago's Brookfield Zoo, where he became a celebrity, attracting more than 300,000 visitors. Harkness brought a second panda, Mei-Mei, in 1938, to be a companion for Su-Lin, but unfortunately they fought and were separated. Su-Lin died at about 2 years old of pneumonia. Su-Lin launched panda mania that still exists to this day. By the early 1950s, the number of pandas in America had dropped to zero; in addition to Su Lin, Mei-Mei and Mei-Lan also died at the Brookfield Zoo; four pandas died at the Bronx Zoo; and the last panda, Pao Pei, died at the Saint Louis Zoo. In 1972, the Chinese government gifted two pandas to President Richard Nixon.

Sugar (Cat)

This cream-colored part-Persian cat with a deformed hip traveled about 1,500 miles for a reunion with his original owners, the Woods. In 1951, they had given him to their neighbors in California before moving to a farm in Oklahoma, because the cat was so terrified of riding in cars. About fourteen months later, Mrs. Woods was in the barn when a cat that looked like Sugar jumped onto her shoulder. At first she didn't believe it was Sugar until she felt the cat's hip. The neighbors in California, who had been given Sugar, later reported that he had vanished two weeks after the Woods left. This case was written about in a parapsychology journal in 1962. There is not a lot of research available on Sugar to confirm this story.

Sun Yat-Sen *see* **Titanic** Animals

Sundance *see* **Tamworth Two**

Surus (Elephant)

The Carthaginian General Hannibal invaded Italy from the north in 218 BCE. He crossed the Alps with the aid of an army of foot soldiers, cavalry, and North African forest elephants. During the trek most of the elephants died, with the exception of a one-tusked elephant named Surus, who became Hannibal's mount. According to Plautus, Surus wore a red cloth and may have also carried a red shield and a construction on his back that served as a platform for Hannibal. Historians believe Surus was an Indian elephant descended from those seized by the Ptolemies of Egypt.

Susan (Dog)

Susan was the first of almost fourteen generations of the more than thirty corgis and dorgis (dachshund/corgi crosses) owned by Queen Elizabeth II throughout her life. The queen's fondness for corgis began in 1944 on her eighteenth birthday, when King George VI gave his daughter,

Princess Elizabeth, this two-month-old Pembroke corgi puppy registered under the name of Hickathrift Pippa; she was called Sue, becoming Susan. A bit of background on the royal legacy of corgis began in 1933 when King George VI (then Duke of York) bought his first Pembroke corgi, named Dookie, from the Rozavel Kennels in Surrey. Seeing that Dookie was welcomed with open arms by his daughters, princesses Elizabeth and Margaret, he purchased a second corgi named Jane. In 1947, Princess Elizabeth wedded Philip, Duke of Edinburgh, at Westminster Abbey. Susan accompanied the couple on their honeymoon as well as on many of their other travels, such as a tour of Earl Mountbatten's Broadlands estate. Susan became the mother of Sugar (Prince Charles' dog) and Honey (the queen mother's favorite) with her first mating with Lucky Strike from Rozavel Kennels. On a later breeding with another Rozavel dog, Rebellion, the queen kept two puppies from the litter. Susan died on January 26, 1959, and was buried in the pet cemetery at Sandringham House. The queen drew up sketches for the gravestone and sent them to superintendent Robert Marrington, with the inscription on the gravestone to read, "Susan/died 26 Jan 1959/for 15 years the faithful companion of the Queen" and later revised to "almost 15 years." Following the deaths of two of her favorites to cancer in 2009, the queen stopped breeding her dogs. At the time of her death, the queen had two corgis, Muick and Sandy.

Swansea Jack (Dog)

Swansea Jack, a black retriever, lived during the 1930s with his owner, William Thomas, near the River Tawe in Swansea, Wales. One day in June 1931, a 12-year-old boy was drowning in the river, and Jack ran in and pulled the boy safely to the shore. However, there were no witnesses to confirm the boy's story. A few weeks later, witnesses observed Jack rescuing another swimmer from the river. Over the next decade, Jack saved at least 27 people from the most dangerous river in Wales. In October 1937, Jack died after eating rat poison.

During his life, Jack was bestowed many honors and awards including:

- A silver collar by the local council in Swansea
- Named the "Bravest Dog of the Year" by *The Star* newspaper in London
- A silver cup from the lord mayor of London
- His own marble and bronze statue
- Named "Dog of the Century" by the National Canine Defence League
- Awarded National Canine Defence League's bronze medal twice
- In 2000 Jack was named "Dog of the Century" by Newfound Friends of Bristol

Sysonby (Horse)

Two-year-old Sysonby, an American Thoroughbred racehorse, took his place in racehorse history with all his victories from 1904 to 1906. Writer Neil Newman ranked Sysonby one of the three best colts, next to Colin and Man o' War, that he had ever seen. From 1904 to 1905, Sysonby had seven major wins and he was the top money earner of 1905, with total career earnings of $184,438. Sysonby died on June 17, 1906, at 4 years and 4 months old, from variola, and in July, James R. Keene donated his remains to the American Museum of Natural History in New York City. Samuel Harmsted Chubb, an anatomist and research associate at the museum, mounted Sysonby's skeleton to show the stride of a running horse. Chubb's skeletons are famous for their studies in anatomy and locomotion. Sysonby's honors included induction into the National Museum of Racing and Hall of Fame in Saratoga Springs, New York, in 1956.

Blood-Horse magazine ranked Sysonby 30th in the list of the top 100 U.S. Thoroughbred champions of the 20th century. His trainer, James Rowe Sr., was inducted posthumously into the Hall of Fame.

Tabby (Cat)

Abraham Lincoln was the first president of the United States to bring a cat to the White House. Tabby went down in history as the first cat to roam the White House. It was Lincoln's son Tad who asked if they could bring their cat Tabby along with them to Washington, D.C. Soon after, Lincoln adopted another feline named Dixie, and he was quoted saying that Dixie was "cleverer than his entire Cabinet."

Tama (Cat)

Tama, a female calico cat, was born on April 29, 1999, in Kinokawa, Wakayama, Japan, and lived with a group of stray cats near Kishi Station. Her claim to fame was for being a station master and operating officer at the Kishi Station on the Kishigawa Line and also for contributing greatly to the growth of the local economy. Toshiko Koyama, the informal station manager, adopted Tama in 2004, when the station was almost shut down due to financial problems. In April 2006, to cut costs the Wakayama Electric Railway replaced all staff on the stations of the Kishigawa Line with employees of local businesses, and that was when Koyama became the official station manager. On January 5, 2007, railway officials named Tama the station master, and with her station master's hat on she greeted passengers. It was estimated that Tama's publicity contributed 1.1 billion yen to the local economy. In January 2008, Tama was promoted to "super station master," a title that came with an office (a converted ticket booth). In January 2010, Tama was promoted again, to "operating officer," becoming the first cat to be an executive of a railroad corporation. Tama's two feline assistant stationmasters were her sister, Chibi, and her mother, Miiko. In Spring 2009, the Wakayama Electric Railway introduced a new train on the line featuring cartoon images of Tama. In August 2010, the Kishi station was refurbished, with the new structure resembling the face of a cat. On June 22, 2015, Tama died of heart failure at age 16. She had a Shinto-style funeral and was given the posthumous title "Honorary Eternal Stationmaster." Next to the station, there is a plaque and a bronze statue of Tama in a small Shinto shrine, called Tama Jinja. Tama's story was part of a documentary about cats titled *La Voie du Chat* in French and *Katzenlektionen* in German, by Italian filmmaker Myriam Tonelotto. The documentary was broadcast on European TV in April 2009. Tama was also featured on the Animal Planet series *Must Love Cats*.

Tamworth Two (Pigs)

On January 8, 1998, these five-month-old brother and sister pigs became a media sensation after escaping slaughter while being unloaded from a truck at the V & G Newman's slaughterhouse in the English town of Malmesbury, Wiltshire. They managed to squeeze through a fence and swim across the River Avon into nearby gardens. The pigs, named Butch (a sow) and Sundance (a boar) after *Butch Cassidy and the Sundance Kid*, were on the run for a week. Their escape piqued the interest of the media and reporters from all over came to cover the story. Filming for the Friday night *NBC News*, reporter Donatella Lorch commented, "These pigs have become celebrities [...] The British reaction to the whole thing is what caught our attention and after all, we are the makers of *Babe*." Despite all the attention, their owner Arnoldo Dijulio still intended to send them to slaughter if or when they were recaptured. In order to save the pigs, the *Daily Mail* newspaper bought the pigs for exclusive rights to their story. Butch (the sister) was captured on January 15 when the two were seen in Harold and Mary Clarke's gardens. The next day, Sundance (the brother) was chased out of a thicket by two springer spaniels. The two pigs lived out their lives at the Rare Breeds Centre, an animal sanctuary near Ashford in Kent. After becoming seriously ill, Butch was euthanized on October 8, 2010. Then seven months later, Sundance, suffering from severe arthritis, was also euthanized at the age of 14 on May 23, 2011. In 2003, the BBC produced a 60-minute television drama titled *The Legend of the Tamworth Two*.

Tangye (Dog)

Tangye was a small black Labrador that made himself at home among the soldiers based at Kajaki, Afghanistan. He lived at the base after a British soldier bought him as a puppy from a dam worker. He began patrolling alongside the soldiers on their campaigns. He survived gun battles and explosions, but the troops feared he was at increased risk from improvised explosive devices (IED). He had been a morale-booster for the soldiers, but the war had become too dangerous to keep him.

Tank and Muck (Dogs)

Tank, an Australian rottweiler, and Muck, a Staffordshire terrier, were both awarded the Purple Cross for their bravery in saving the life of two-year-old Max Hillier in 2008. The Hillier family was visiting a friend at Andergrove near Mackay, Queensland, and they brought along their dog Tank. Max, Tank, and their friend's dog Muck were playing in the backyard and wandered off into the woods, ending up near a dam. Max fell into the dam and the two dogs leaped into action and managed to drag Max out of the water to safety. Upon hearing frantic barking, the neighbor rushed out to find Tank guarding Max and Muck pacing between the boy and the ledge of the dam.

Taro and Jiro (Dogs)

Taro and Jiro were two three-year-old male Sakhalin husky siblings. Starting in January 1957, they were the two youngest of 15 sled dogs leading the first team of 11 researchers on a Japanese expedition to Antarctica. In February 1958, a second team of researchers was to be sent to replace the first team, but an unexpected storm caused the expedition to be suspended, and the first team was evacuated by helicopter. The dogs had to be left behind; with the researchers hoping to return soon, they were chained and given food to last a few days. Sadly, the men were not able to return and on January 14, 1959, the expedition returned to base camp expecting to find all 15 dogs dead. However, only seven (Aka, Goro, Pochi, Moku, Kuro, Pesu, and Kuma) died while still chained to the fence. Six other dogs (Riki, Anko, Deri, Jakku, Shiro, and a second dog named Kuma, who was the father of Taro and Jiro) were never found. Only Taro and Jiro survived over 11 months in extreme weather and, without the care of humans, did not resort to cannibalism. Professor Emeritus Yasukazu Kitamura of Kyushu University described the situation: "The stored foods for human and dog in Showa base were intact. Some types of dogs have to cannibalism [sic] during the famine. Seven dogs were found dead which were connected to chains, while the bodies were intact. I think that Sakhalin dogs ate penguins, feces of seals, seabirds and fishes (trapped in the ice)." In 1960, Jiro died of disease in Antarctica; his body was embalmed and is on display at the National Museum of Nature and Science in Tokyo. Taro died of old age in 1970,

and his body was also embalmed and is on display at the Museum of National Treasures at the Botanical Garden of Hokkaido University. There are monuments in Japan honoring Taro, Jiro and all the dogs that died. There are also two movies based on their lives, *Nankyoku Monogatari* (1983) and *Eight Below* (2006).

Tarzan (Horse)

Tarzan starred with silent and sound film actor Ken Maynard in five dozen films and serials from 1925 until Tarzan died in 1940. Unlike other "wonder horses" that were known for their daring stunts, Tarzan was better known for his tricks, which included dancing, bowing, answering questions by nodding his head, playing dead, untying ropes, and nudging Maynard into the arms of the leading lady. Tarzan was so intelligent that he was able to perform these tricks simply from Maynard's spoken commands.

Taz (Dog)

"Taz the Wonder Dog," a mixed breed with a remarkable story, is the canine hero in Bob Babbitt's book *Never a Bad Day*, about the bond between Danelle Ballengee and her dog Taz. Danelle was a premier athlete winning or participating in many outdoor competitions, not to mention winning six U.S. Athlete of the Year Awards in four different sports. This particular day started out to be just like any other workout but would turn into a fight for survival. Danelle and Taz, which she had owned for three years, set out for a two-hour jog from a trailhead just a few miles from her house in Moab, Utah. Along her route, she slid on some black ice and plunged 60 feet, landing on a rock shelf, badly injured. Unfortunately, she had not told anyone where she was going and she was not dressed for temperatures in the low 20s. On the second night, Danelle's parents and her neighbor reported her missing, and on the third day a search and rescue team was sent out. On the same day, Taz began running to the trailhead, about five miles away, to look for help. The rescue team found Danelle's truck and Taz; the dog led them right to Danelle. Taz received a Christmas stocking from Michigan containing a six-pack of steaks—a happy ending for all.

Teddy (Dog)

In October 1943, Teddy's War Dog Unit was assigned to a Marine Raider Regiment of the Sixth Army. In December, Teddy and his handler moved with the regiment to Finschhafen to be a part of the Cape Gloucester operations in the South Pacific. Teddy's unit was assigned to the Army Forces near Gilnit, and none of the dogs failed their missions, nor were their patrols ambushed or casualties suffered. The patrols led by dogs were credited with 180 Japanese casualties and 20 prisoners.

Teka (Dog)

Teka, an Australian cattle dog, received the RSPCA Animal Achievement Award for saving her owner Jim Touzeau's life in 2007. Jim collapsed from a heart attack on the floor in a glass factory in Tinana, Queensland, and Teka ran to his side. Teka startled Jim awake by pouncing on his chest with her two front paws and barking in his face. Outside she attracted attention by jumping and barking on the street. Did Teka simulate CPR to revive Jim's heart? Michael Betty of the RSPCA and some medical experts debated this question; in any event according to doctors, the fact that she forced Jim to wake up saved his life.

Terrible Ted (Bear)

Terrible Ted, a Canadian American black bear, was a professional wrestler from the 1950s to the 1970s. "Ted the Wrestling Bear" made his debut on April 1, 1950. After his debut, he wrestled four matches. However, his trainer and manager, Dave McKigney, speculated that these matches may have involved another bear, since in 1950 Ted would only have been six months old and have weighed less than 100 pounds. In 1959, *The Regina Leader-Post* reported that in Ted's five-year career thus far he had won over 500 matches. Ted continued winning matches until he retired in January 1974 after defeating Sam Bass in Nashville. In his early years, Ted had traveled with a carnival and, declawed and detoothed, was adopted by McKigney after the carnival went bankrupt; on McKigney's property in Aurora, Ontario, Ted joined another bear, Smokey. On July 2, 1978, Smokey got into McKigney's house and mauled his girlfriend, Lynn Oser, to death. Both bears were taken by the Ontario Humane Society. McKigney's explanation was that at times bears could be unpredictable during mating season. The bears' fate is uncertain.

Terry (Dog)

Terry was a female Cairn terrier actress born on November 17, 1933, and owned and trained by Carl Spitz. Terry is most famous for her role as Toto in the film *The Wizard of Oz* (1939). Other film appearances include *Ready for Love* (1934), *Bright Eyes* (1934), *The Women* (1939), *Bad Little Angel* (1939) and her last film, *Tortilla Flat* (1942). She had a total of sixteen film appearances. During the filming of *The Wizard of Oz* her foot was broken by one of the Wicked Witch's soldiers, and she had to recuperate for over two weeks. In 1942, her name was officially changed to Toto. Terry died on September 1, 1945, at age eleven.

Other films include:

The Dark Angel (1935)
Fury (1936)
The Buccaneer (1938)

Stablemates (1938)
Calling Philo Vance (1940)
Son of the Navy (1940)
Cinderella's Feller (1940)
The Chocolate Soldier (1941)
Twin Beds (1942)
George Washington Slept Here (1942)

That Devil Dan *see* **Daniel Webster**

Theo (Dog)

Theo, a male English springer spaniel born in May 2009, served as a bomb-sniffing dog with the British Army in Afghanistan. In September 2010, Theo and his handler, Lance Corporal Liam Tasker, were sent on their first tour of duty to Afghanistan. Over five months, Tasker and Theo found fourteen bombs. While on patrol with the 1st Irish Guards in the Nahri Saraj District in Helmand Province on March 1, 2011, Tasker was killed by a sniper. Just a few hours after Tasker's death, Theo died from a seizure. Tasker's mother, Jane Duffy, thought that Theo had died of a broken heart. Theo was the sixth dog to die in Afghanistan. On October 25, 2012, Theo was posthumously awarded the Dickin Medal.

Thibault (Lobster)

The 19th century poet Gerard de Nerval was reportedly seen walking his lobster around the gardens at the Palais-Royal on a blue silk leash. In response to some curious passerby, he explained, "Why should a lobster be any more ridiculous than a dog? Or a cat, or a gazelle, or a lion, or any other animal that one chooses to take for a walk? I have a liking for lobsters. They are peaceful, serious creatures. They know the secrets of the sea, they don't bark, and they don't gobble up your monadic privacy like dogs do."

Thumbelina (Horse)

This miniature chestnut mare, at 17 pounds, was certified the world's smallest horse by the Guinness World Records in 2006 (the record is now held by Bombel, a miniature Appaloosa from Poland). Thumbelina was born on May 1, 2002, on a farm in St. Louis, Missouri. Because both of her parents were standard miniature horses, standing 30 inches tall, it was quite a shock when at birth she weighed 8 pounds and stood 10 inches tall. Despite her fragile condition at birth, she overcame some of her early struggles such as her crooked legs, which required special shoes, and she ran, bucked and played. In just one year Thumbelina became one of the most famous animals in history; she was featured in thousands of newspapers and magazines and appeared in hundreds of television broadcasts all over the world. In the United States she appeared on: *The Oprah Winfrey Show, Live with Regis and Kelly, Inside Edition, The Today Show, Good Morning America, The Early Show, Fox & Friends,* and on The Discovery Channel. Thumbelina had a special connection with children, who thought she was magical. In 2007, the Thumbelina Charitable Foundation was established to raise awareness and funding for children's charities. In April 2007, the Thumbelina Children's Tour was launched to visit children at hospitals, shelters, group homes, military bases, special schools and camps. Michael Goessling left his career as a marketing consultant to be Thumbelina's owner and handler. He and Thumbelina toured North America in the "Thumbymobile" visiting children in need. Thumbelina died in 2018.

Tibbles (Cat)

Lyall's wrens, or Stephens Island wrens, are an extinct species of small, flightless, nocturnal birds with olive-brown plumage. They were widespread throughout New Zealand until the settlement of the Maori, and by 1894 were only found on Stephens Island. By 1894, a lighthouse had been built, and June 1879 marks the first significant human activity on Stephens Island. The birds were first spotted by David Lyall, assistant lighthouse keeper. His cat, Tibbles, would bring him dead bodies of the birds that nobody knew existed. Though Tibbles is often blamed for the extinction of the Lyall's wren, there were other feral cats on the island that hunted the flightless bird for food. Cats are not native to New Zealand. Europeans who arrived there had pregnant cats escape, resulting in the increase of cat populations as an invasive species. In an attempt to rid the island of feral cats, authorities killed many of them. It took 26 years but by 1925 the island was declared free of cats.

Tiger (Cat)

Tiger was a tom cat that Calvin Coolidge moved into the White House from his farm in Vermont. He would come when called by his nickname "Tige," and Coolidge liked to walk around with the cat draped around his neck. On the night of March 20, 1924, Tiger slipped out an open door and disappeared, and when Coolidge called for Tiger he did not come. The president had the staff search the White House and grounds. He asked the city police to be on alert for the orange and black cat. He sent a Secret Service agent to WCAP radio to broadcast an appeal to listeners. After the station had received hundreds of calls, one listener finally called and said he had found a sleeping cat in the Navy Building that responded to "Here, Tige." Tiger was soon back home. Tiger got a new collar with the inscription, "My Name is Tiger. I live at 1600 Pennsylvania Avenue." He strayed a second time and was never seen again.

Tiger (Lion)

Marguerite Durand, a feminist icon, was a French stage actress, journalist, and women's rights advocate. She was often seen around Paris with her exotic pet lion, Tiger. She was also co-founder of the Cimetière des Chiens et Autres Animaux Domestiques (Cemetery of Dogs and Other Domestic Animals) in Paris. Tiger is buried in this cemetery due to Durand's use of the word *autres*, or "other," in the cemetery's name.

Timothy (Tortoise)

In 1854, the British navy discovered a tortoise and named it Timothy. Timothy became a mascot for many ships at sea. During the Crimean War, Timothy served on the HMS *Queen*. After serving many years at sea, "he" retired to live at Powderham Castle Rose Garden. In 1926, it was discovered that Timothy was actually a female tortoise. During World War II, when she felt the vibrations from nearby bombings, she would begin to dig her own little shelter. Timothy died in 2004 and was believed to be around 165 years old.

Tinker (Dog)

Tinker was considered a valuable crew member and a good luck charm by Captain Michael Philip Usina of the Confederate blockade runner *Armstrong* and his crew. The charm was not a rabbit's foot, but a living four-legged terrier named Tinker. Tinker served as crew mascot and became a valuable asset with his ability to detect the enemy or sense trouble. Some sailors refused to sail without Tinker on board. One rival captain offered Usina, who was Tinker's owner, $500 in gold for just one trip with Tinker. Usina refused. Tinker was buried at sea in the North Atlantic in April 1865, during Usina's voyage to England.

Tirpitz (Pig)

The Times newspaper reported:

The animal, which is known as "Tirpitz," was once owned by the German light cruiser *Dresden*, and when, during the action with *Glasgow, Kent,* and *Orama,* the Germans escaped to the shore after causing an explosion which sank the *Dresden,* and "Tirpitz" was left to its fate, the pig struck out boldly, and was seen swimming near the *Glasgow.* Two sailors dived into the sea, and the animal was brought safely aboard. The ship's company of the *Glasgow* awarded "Tirpitz" an "Iron Cross" for having remained in the ship after its shipmates had left, and it became a great pet.

The *Dresden* was defeated at the Battle of the Falkland Islands, but managed to escape and on March 15, 1915, was found in Cumberland Bay on the Chilean island of Mas a Tierra (today known as Robinson Crusoe Island) by HMS *Glasgow* and HMS *Kent.* After being rescued, Tirpitz was adopted by the *Glasgow* crew and made their mascot—naming her Tirpitz (some accounts say the pig was male) after Alfred von Tirpitz, the German admiral and secretary of state of the Imperial Naval Office. Tirpitz was adopted by the petty officer who was the first to see her in the water and spent the rest of her career at Whale Island Gunnery School in Portsmouth. In 1919, to raise money for the British Red Cross, she was auctioned off as pork. William Cavendish-Bentinck, 6th Duke of Portland, donated Tirpitz's stuffed head to the Imperial War Museum. In 2006, the head was part of the museum's temporary exhibit The Animals' War.

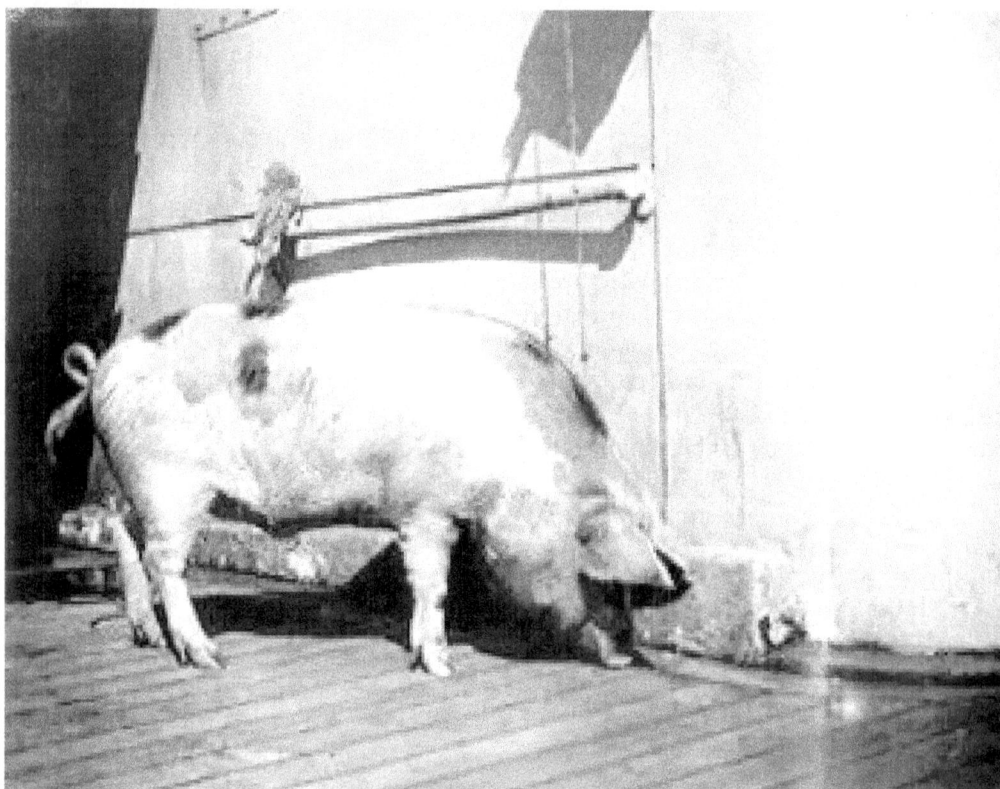

Following the defeat of the German light cruiser *Dresden*, Tirpitz (shown here in May 1915) was rescued from the sea by two sailors on March 15, 1915. She became the mascot of their ships HMS *Glasgow* and HMS *Kent* (Royal Navy official photographer, Imperial War Museum catalogue number Q 47559).

Titanic Animals

At least a dozen *Titanic* passengers have received less attention than their human counterparts—roughly twelve dogs were on board the *Titanic* on April 15, 1912—all pets of first-class passengers. Three dogs survived—two Pomeranians and a Pekingese. They made it alive due to their size; they were so small that it's doubtful anyone even realized they were on the lifeboats. There may have been even more dogs on board than known. Some of the *Titanic's* canine casualties included a fox terrier named Dog, an Airedale named Kitty, and a French bulldog named Gamin de Pycombe. One passenger refused to leave without her Great Dane and later both were found floating at sea.

1. Jenny, the ship's cat, was seen leaving the ship carrying her kittens off before it sailed.

2. Lady, a Pomeranian that had been purchased in Paris by Margaret Bechstein Hays.

Another Pomeranian, whose name is unknown, was owned by New York clothing magnate Martin Rothschild and his wife Elizabeth Jane Anne. While Martin did not survive, his wife with her hidden dog made it to a lifeboat.

3. Sun Yat-Sen, a Pekingese, was owned by Henry S. Harper and his wife Myra. The Harpers were returning from a tour of Europe and Asia.

Some passengers who left their pets received some consolation in the form of insurance payments. There are stories of other animals on the *Titanic*, but none are confirmed.

Toby *see* **Black Diamond**

Toby (Pig)

The "Learned Pig" Toby was trained by Scotsman Samuel Bisset, who ran a traveling novelty show in 1780s London.

Toby was taught to pick up cards with his mouth to answer arithmetic problems and spell out words—causing quite a sensation. During his four-year career, Toby toured a great deal, performing in Dublin, Nottingham, London, and then on to continental Europe. Performing horses were not unusual, but it was only after Toby that other trained pigs became features in fairs and other attractions throughout Europe and America in the 19th century. Even though the "Learned Pig" shows were great successes, there was also some debate as to how the pig was trained; some believed that some sort of torture may have been involved. A children's book by Sarah Trimmer titled *The Robin Redbreasts* (1788), was written to teach children how to properly treat animals. Toby inspired and was featured in both caricatures and literature. In 1786 a pamphlet, *The Story of the Learned Pig*, by an anonymous officer of the Royal Navy, examined the topic of reincarnation.

Thomas Hood's poem "The Lament of Toby, The Learned Pig" describes a learned pig's thoughts of retirement leading to slaughter:

> In this world, pigs, as well as men,
> Must dance to fortune's fiddlings,
> But must I give the classics up,
> For barley-meal and middlings?

William Wordsworth's poem "The Prelude" refers to the learned pig as one of the "freaks of nature":

> …Albinos, painted Indians, Dwarfs,
> The Horse of knowledge, and the learned Pig
> All out-o'-the-way, far-fetched, perverted things,
> All freaks of nature, all Promethean thoughts
> Of man, his dullness, madness, and their feats
> All jumbled up together, to compose
> A Parliament of Monsters.

In his book *The Mudfog Papers* Charles Dickens describes a lecture given by Mr. Blunderum at the Mudfog Society for the Advancement of Everything on the learned pig's dying thoughts.

Also in more recent times "The

Learned Pig" appears in popular culture. In 2003, the UK band Tiger Lillies recorded a song, "The Learned Pig," based on a poem by Edward Gorey. In 2007, a musical, *Toby the Incredible Learned Pig* written by Daniel Freedman, was a finalist at the WonderLand One-Act Festival at TheaterWorks in New York City. A 2011 novel, *Pyg: The Memoirs of a Learned Pig* by Russell Potter, is based on Toby's career.

Togo (Dog)

Togo was born on October 17, 1913, and was euthanized on December 5, 1929, at the age of 16. The offspring of one of Leonhard Seppala's former lead dogs, Suggen, Togo did not start out to be a sled dog. Topping out at 48 pounds at 6 months old, he was considered too small by comparison to other sled dogs. Plus he was seen as "difficult and mischievous," showing "all the signs of becoming a … canine delinquent," according to one reporter. His bad behavior was thought to be because he was spoiled by Seppala's wife when he was sick as a puppy. Seppala did not see Togo sled-dog bound, so he gave him away as a pet at six months of age. A few weeks later, Togo proved his loyalty to the team by running several miles back to his original owner, impressing Seppala so much that he never gave him away again. However, Togo still continued his mischievous ways by attacking the lead dogs of oncoming teams. One day he learned a valuable lesson when he was severely injured from attacking a malamute leader, teaching him to keep a wide berth with oncoming teams. When Togo was 8 months old Seppala put him in a harness to try to control his behavior, and surprisingly Togo immediately settled down and was calm. On his first day in the harness, Togo logged an unheard of 75 miles. Thus Seppala kept moving him up the line until eventually Togo was sharing the lead with Rusky. At age 12, Togo ran with Seppala on part of the

historic "Serum Run," traveling 170 miles from Nome, Alaska, to pick up the first batch of 300,240 units of lifesaving serum in Shaktoolik on January 31. The journey was made under very severe conditions—temperatures at minus 30 degrees Fahrenheit and below, gale force winds, and exposed open ice. The team climbed 5,000 feet to cross Little McKinley Mountain and then descended to Golvin, passing the serum to Charlie Olsen, who then passed the serum to Gunnar Kaasen and Balto. Balto was the dog credited with saving the town, but Togo's journey was longer by 200 miles. In October 1926, Seppala took Togo and a team of dogs on tour from Seattle, Washington, to California, drawing large crowds in stadiums and department stores. In New York City, the team walked along Fifth Avenue and through Central Park and Madison Square Garden. Togo won several awards during his lifetime (Most Traveled Dog in Alaska, Champion Trophy Winner in Nome, Record of Longest and Fastest Run in Serum Drive). Togo retired in Poland Spring, Maine. When Togo was a puppy Seppala called him an "infant prodigy" and added, "I had found a natural-born leader, something I had tried for years to breed." Historian Earl Aversano reported that in his old age, Seppala recalled, "I never had a better dog than Togo. His stamina, loyalty and intelligence could not be improved upon. Togo was the best dog that ever traveled the Alaska trail." After Togo's death, his mounted skin was displayed at the Shelburne Museum in Vermont. Today his mounted skin is at the Iditarod Trail Sled Dog Race Headquarters museum in Wasilla, Alaska. His skeleton is in the collection at the Peabody Museum of Natural History at Yale University.

Tony (Horse)

Between 1922 and 1932, Tony appeared in thirty-four films and was the first horse to be nicknamed "The Wonder Horse."

Tony was the first horse to share equal billing with his co-star, Tom Mix, and also had his name appear in three movies: *Just Tony* (1922), based on the short story "Alcatraz" by Max Brand, *Oh! You Tony* (1924) and *Tony Runs Wild* (1926). Tony was a true celebrity; in 1927, the horse's hoofprints were placed alongside Tom Mix's handprints in the cement at Grauman's Chinese Theater. Tony is well-known for his stunts, which were performed before the American Humane Society began overseeing the treatment of animals in film productions. Some such dangerous stunts included: Mix and Tony jumping through a glass window and running alongside a speeding train in the 1926 film *The Great K & A Train Robbery*; and in the 1921 film *Trailin'* a bridge collapses underneath Tony and Mix, and some uncut film shows the two falling into the river below. There was some suspicion that there may have been some abuse involved in creating some of the stunts. Due to an injury while filming his last movie, *The Fourth Horseman* (1932), Tony retired at the age of twenty-three and died ten years later. Tony's death was published in *The New York Times*. Tony outlived Tom Mix (who died in a car accident) by two years. Tony is also remembered for being featured in junior novels, comic books and a 1934 children's book, *Tony and His Pals*.

Top (Wombat)

Wombats are nocturnal hairy creatures native to parts of Australia. Victorian Europe's trade in exotic animals thrived, and even though kangaroos were more popular, wombats were rarer and more expensive. Napoleon and the Duke of Edinburgh both owned wombats, but this is the story of Top, one of two pet wombats belonging to the pre-Raphaelite painter and poet Dante Gabriel Rossetti. In 1862, Rossetti moved to Tudor House at 16 Cheyne Walk in Chelsea. The property had four-fifths of an acre of garden,

which he filled with exotic birds and animals and in September 1869 added the wombat. Top arrived when Rossetti was recovering from a breakdown in Scotland, where he had formed an infatuation with Jane Morris, the wife of an old friend and protégé. During this time, Rossetti drew an illustration of Jane and the wombat, both with halos, showing that in his mind the two were objects of sanctification. Upon returning to London, Rossetti wrote to William Michael his most famous remark about Top: "The wombat is a joy, a triumph, a delight, a madness." Top died on November 6, 1869, shortly after he was purchased, from some kind of mange-like illness. After Top's death, Rossetti stuffed and displayed him in the front hall. Some silly stories emerged about Top: James McNeill Whistler invented the story that Top died after consuming a box of cigars; Ford Madox Brown wondered if Rossetti's habit of dining with Top and allowing him to sleep on the dining room table may have inspired Lewis Carroll's "dormouse in the teapot incident" at the Mad Hatter's tea party, which was impossible; Lewis Carroll wrote that chapter in 1863 and the novel was published in 1865— well before Top's purchase by Rossetti and subsequent death. Rossetti turned to the Canadian woodchuck in place of his two pet wombats.

Topsy (Elephant)

Topsy was a female Asian elephant born in Southeast Asia around 1875. Topsy was named after a slave-girl character in *Uncle Tom's Cabin*. Shortly after her birth, she was captured by elephant traders and smuggled into the United States by Adam Forepaugh, owner of the Forepaugh Circus. Being in competition with the Barnum & Bailey Circus, he planned to advertise Topsy as the first baby elephant born in America, which he did in 1870. The very same elephant traders tipped Barnum off about the hoax, and after being exposed, Forepaugh advertised her as the first elephant born outside of a tropical zone. Full-grown, Topsy stood 10 feet tall, 20 feet long and weighed between 4 and 6 tons. Over the 25 years that she was with Forepaugh, she gained a reputation as a "bad" elephant, especially after killing spectator James Fielding Blount in Brooklyn, New York, in 1902. It was believed that on the morning of May 27, Blount was drunk when he wandered into the elephant's tent, and he threw sand in Topsy's face and burnt the tip of her trunk with a cigar. Another attack took place in June 1902 when Topsy raised spectator Louis Dodero in the air with her trunk and threw him down after he tickled her behind the ear. The American Society for the Prevention of Cruelty to Animals stopped Luna Park management's plans to hang Topsy and charge admission. However on January 4, 1903, in front of a small group of press and invited guests, Topsy was electrocuted and strangled. It was the electrocution that killed her. To prepare her, the executioners fed her carrots laced with more than a pound of potassium cyanide and strapped on metal shoes; then they were ready to zap her with 6,600 volts. Fortunately it only took seconds for Topsy to die. The Edison Manufacturing Company's movie studios released a film, *Electrocuting an Elephant*. Some reports say Topsy's execution was a "key moment" in the so-called "Battle of the Currents," while other reports describe Thomas Edison's role in the execution as a myth. Edison was having a great deal of difficulty proving that his model of direct-current electricity was safer and more efficient than alternating current (AC). Being a ruthless businessman, he decided that the best way to show the American public that his model was superior to the AC technology was to electrocute defenseless animals with the AC technology to demonstrate its danger. Some reports say Topsy was chosen to be yet another experimental subject, while other reports say the elephant's execution had already been decided by Luna Park management. Edison's film was shown in theaters worldwide.

Toto *see* **Terry**

Trakr (Dog)

Trakr was trained in the Czech Republic and joined the Halifax Regional Police in Canada when he was fourteen months old. Trakr's handler, a Canadian police officer, saw what had happened at the World Trade Center on September 11, 2001, and the two left for New York the next day. Trakr searched through the debris and rescued a woman who had been trapped in the rubble for twenty-six hours and was the last person found alive. Trakr worked at Ground Zero until September 14, when he collapsed; he returned to Canada to recover. Trakr passed away in April 2009 and months later, five puppies were cloned using his DNA.

Traveller (Horse)

Traveller was born and raised by Andrew Johnston near Blue Sulphur Springs in Greenbrier County, Virginia (now West Virginia). This 4-year-old gray gelding of American Saddlebred breeding, originally named Jeff Davis and then Greenbrier, was one of the many famous horses from the Civil War. Before he was famous he won prizes at county fairs in 1859 and 1860. In the spring of 1861, Captain Joseph M. Broun, the quartermaster of the 3rd Regiment, Wise Legion, was directed to "purchase a good serviceable horse of the best Greenbrier stock for our use during the war." Broun purchased the horse for $175 and named him Greenbrier. The horse was then purchased from Broun for $200 by Confederate General Robert E. Lee in February 1862, and the

Traveller was purchased in February 1862 by Confederate general Robert E. Lee and the two were companions until Lee's death (photograph by Michael Miller, in Francis Trevelyan Miller and Robert S. Lanier, *The Photographic History of the Civil War* [New York: Review of Reviews Col., 1911]).

two were together until Lee's death. Lee soon renamed him Traveller and used him almost daily for the duration of the Civil War. Traveller had great stamina and was difficult to frighten. However at the Second Battle of Bull Run, while General Lee was leading him by the bridle, Traveller got spooked and pulled Lee down on a stump, breaking both of the general's hands. After surrendering to Ulysses S. Grant at the Appomattox Courthouse in Virginia on April 9, 1865, Lee mounted Traveller and road back to

his troops. When Lee became president of Washington College in Virginia, he road Traveller on a 108-mile trip to Virginia. After a long life Traveller was put down in 1871 after contracting tetanus when he stepped on a nail while walking behind Lee's hearse. After the horse's death his remains were moved from one place to another until 1971, when they were buried next to University Chapel (formerly Lee Chapel) on the Washington and Lee University campus. Traditionally his stable doors were left open so his spirit could wander freely. Thomas Burish, the 24th president of Washington and Lee, was strongly criticized for closing the stable gates, and later had the doors repainted a dark green and referred to the color as "Traveller Green." The base newspaper of the United States Army's Fort Lee in Petersburg, Virginia, was called the *Fort Lee Traveller*; its publication ended in 2021.

Passage from "Army of Northern Virginia," a poem by Stephen Vincent Benét:

And now at last,
Comes Traveller and his master. Look at
  them well.
The horse is an iron-grey, sixteen hands
  high,
Short back, deep chest, strong haunch,
  flat legs, small head,
Delicate ear, quick eye, black mane and
  tail,
Wise brain, obedient mouth.
                                    Such horses are
The jewels of the horseman's hands and
  thighs,
They go by the word and hardly need the
  rein.
They bred such horses in Virginia then,
Horses that were remembered after death
And buried not so far from Christian ground
That if their sleeping riders should arise
They could not witch them from the earth again
And ride a printless course along the grass
With the old manage and light ease of hand.

Passage from *Traveller*, a novel by Richard Adams:

Their sleepless, bloodshot eyes were turned to me.
Their flags hung black against the pelting sky.
Their jests and curses echoed whisperingly,
As though from long-lost years of sorrow–Why,
You're weeping! What, then? What more did you
  see?
A gray man on a gray horse rode by.

Trent *see* **Kiddo**

Treo (Dog)

Treo was a black Labrador retriever and English springer spaniel mix, and an Arms and Explosive Search dog with the Royal Army Veterinary Corps. Treo was donated to the Royal Army because of his inclination for snapping and growling at people. After a twelve-week training course he was deployed to Northern Ireland for three years. In 2008, Treo and his handler were deployed to Afghanistan to support British troops there. He had great success finding IEDs and saving the lives of British soldiers. He was awarded the Dickin Medal in 2010. After seven years in the military, he retired and lived with his handler, Sergeant Heyhoe. He died in 2015 and was buried with his Dickin Medal.

Trigger *see* **Golden Cloud**

Trim (Cat)

Trim, a black cat with white paws and chest, was born in 1799 on board the ship HMS *Reliance*, commanded by Matthew Flinders. Trim was named after the butler who went by the name Trim in Laurence Sterne's novel *Tristram Shandy*. Between 1801 and 1803, Trim, the "ship's cat," accompanied Flinders and his crew on their voyages to circumnavigate the coastline of Australia. As a kitten Trim fell overboard on their voyage from the Cape of Good Hope to Botany Bay. To the surprise of the crew, he managed to swim back to the ship and scaled a rope to climb aboard. Trim also accompanied Flinders' circumnavigation of the Australian mainland on the HMS *Investigator*. Together Trim and Flinders survived a shipwreck on Wreck Reef in 1803 and were captured by the French. When returning to England, Flinders was accused of spying and imprisoned in Mauritius. During this time Trim mysteriously disappeared and Flinders assumed he may have been stolen and eaten by a slave. There have been a number of tributes to both Flinders and Trim. In 1996, a bronze statue of Trim was erected behind a statue of Flinders on the

This bronze statue of Trim sits on the window ledge of the Mitchell Library in Sydney, Australia (CC BY-SA 3.0).

window ledge of the Mitchell Library in Sydney. Trim is also found on a statue of Flinders in Donington, England. There are also a two statues of Trim at the foot of Flinders: one in Port Lincoln, South Australia, and the other at Flinders University in Adelaide, South Australia. At the Euston Railway Station in London there stands a large bronze statue of the two. There is also a novel, *Matthew Flinders' Cat*, by Australian author Bryce Courtenay. Lastly, the café at the State Library of New South Wales is named after Trim, and a plaque under his statue outside the library reads:

TO THE MEMORY OF TRIM

The best and most illustrious of his race
The most affectionate of friends,
faithful of servants,
and best of creatures
He made the tour of the globe, and a voyage
to Australia, which he circumnavigated, and
was ever the delight and pleasure of his fellow
voyagers.

Written by Matthew Flinders in memory of his cat. Memorial donated by the North Shore Historical Society.

Trouve (Dog)

When Alexander Graham Bell was twenty years old, he taught Trouve, a rescued stray Skye terrier, how to "talk" by manually manipulating the dog's lips and vocal cords to produce a variety of sounds. Using methods created by his father, Bell would shape the dog's lips to produce the sound "ma" and taught Trouve to say "mama" every time he begged for a treat. One sentence that Trouve could "say" was, "How are you, Grandma?" Bell proved that his methods could be used to teach deaf people to talk.

Tsekoff (Dog)

Tsekoff was a guard dog, and on his first night of duty he found two spies hiding out in a flour wagon and attacked them.

Tusko (Elephant)

Tusko, formerly known as Ned, was captured at six years old in Siam (now Thailand) and was unloaded at New York Harbor in 1898. By 1922, he was known as the "The Meanest Elephant" and also "the largest elephant ever in captivity." His tusks were seven feet long, thus the name "Tusko." On May 15, 1922, when the Al G. Barnes Circus came to the Skagit Valley logging town of Sedro-Woolley, no one was prepared for Tusko's wild romp through town. Tusko initially stumbled about the circus grounds, taking off north and uprooting trees, knocking down telephone poles, tearing down fences, leaving his huge footprints on lawns, and destroying a chicken coop. About 9:00 the next morning, a posse successfully corralled Tusko. No one was hurt during Tusko's adventures; however Al Barnes had to pay $20,000 in damages. Not wanted in any circus, and after a series of different owners, Tusko ended up transferred to the Woodland Park Zoo in Seattle on October 8, 1932. Children created a special fund to pay for his food and the zoo hired famous trainer George "Slim" Lewis, paying him $3.50 per day to tend to the animal in a specially enlarged elephant house. Tusko died on June 10, 1933, of a blood clot in his heart. In the 1950s, Tusko's skin and bones were donated to the University of Oregon's Museum of Natural and Cultural History.

Tusko (Elephant)

This male Indian elephant was kept at the Oklahoma City Zoo. He was part of a research study of LSD (lysergic acid diethylamide) at the University of Oklahoma. On August 3, 1962, researchers administered 297 mg of LSD to Tusko—1,000 times the dosage for human recreational use.

Five minutes later, he collapsed and one hour and forty minutes later he died from the LSD and also perhaps from the drugs used to revive him.

Tusko (Elephant)

Tusko was an Asian elephant born in Thailand in the 1970s, and he was on breeding loan at the Oregon Zoo in Portland from 2005 until he was euthanized on December 22, 2015, due to an infected right front foot. Tusko had successfully sired three calves in the past: two in Canada and one in California. At the Oregon Zoo he mated with Rose-Tu and on August 23, 2008, she gave birth to a male calf named Samudra (nicknamed "Sam"); on November 30, 2012, she also gave birth to a female calf named Lily. Lily was sold to Perris, California-based Have Trunk Will Travel, a company whose business was to offer elephant rides at fairs, zoos and weddings. After some contractual agreements, the Oregon Zoo Foundation held a fundraiser, and Tusko and Lily were purchased by the zoo for $400,000.

Tweed (Dog)

Tweed was a very shy bobtailed sheepdog that, after working with Major Richardson's wife at war-dog school, became one of the most reliable messenger dogs in the British army. Tweed served with a Scottish Canadian regiment at Amiens in 1918. The Germans cut off the British front line and would have captured the town if it were not for three messenger dogs, one being Tweed, that were sent to the French Colonial headquarters with the message, "Send up reinforcements and small round ammunition." Thus Amiens was saved from the Germans.

Two-Bits (Dog)

Two-Bits, the World War II Lookout Dog, was a fox terrier that spent the winter of 1942–1943 with his owner Bill Zeigler and another Aircraft Warning Service man at the Siskiyou Mountains' Whisky Peak Lookout. Before radar the Army Air Corps used U.S. Forest Service fire-lookout facilities on the Pacific Coast to scan the western skies for enemy aircraft. Two-Bits' winter quarters were a 14-by-14 building on the summit next to a vertical 600-foot drop to Low Gap Creek. The team's job was to report to the army any aircraft heard or observed. Their food and supplies were delivered every two weeks by Forest Service members on skis. Two-Bits accidentally plunged over the cliff twice while chasing rodents but managed to survive both falls. After Zeigler and Two-Bits returned to Jacksonville in the summer of 1943, the Medford *Mail Tribune* reported that because Two-Bits had survived his falls "without physical impairment or loss of morale," he was a symbol of Home Front stolidity and determination.

Two Bits (Horse)

Being one horse of a cavalry pool at Fort Craig, New Mexico, Two Bits had known a dozen masters until he was assigned to Captain Charles A. Curtis. Between the 1870s and 1880s the United States was trying to force the Native Americans—mostly Apaches, Comanches, and Navahos—to stay on their reservations. However, the tribes were not happy with the government's methods and constantly raided the forts. It was at Fort Craig where Two Bits got his name. To relieve some of their stress the men arranged a horse race. All of the horses of the Mounted Rifles were chosen by the riders, except a big bay horse. Cain, an Irish fifer boy, chose this horse. At the starting line one soldier exclaimed, "I wouldn't give two bits for that horse." Two Bits won the race by three lengths. Six years later Sergeant Cain, while serving at Fort Whipple in Arizona, came upon Two Bits, nearly starved, with

his brutal master beating him. It seems that Two Bits was among a group of horses no longer fit to serve in the cavalry that were sold at auction. Cain ran to rescue the horse, and to avoid a fight, the owner willingly gave him up. Knowing that the army would not accept the horse and he could not afford to keep him, Cain thought his captain, Curtis, would give Two Bits a good home. A few weeks later Cain arrived with a groomed and filled-out horse, and the captain and horse fell in love with each other. Two Bits really became a hero when he rode with a Sergeant Porter to deliver a crucial dispatch to Santa Fe. After traveling for four days the two ran into a battle and both were badly wounded. Porter fell to the ground and did not respond when Two Bits nudged him; the horse smelled fire, galloped towards the smoke and came upon a government train guard huddled around a fire. The horse attempted to lead the soldiers to the wounded Porter. The soldiers mounted their horses waiting for him to take the lead, but Two Bits staggered forward and died. Sergeant Porter was rescued and recovered from his wounds. Two Bits was given a soldier's burial and laid to rest under shady pines along with twenty-one cavalrymen.

Two Feathery Soldiers (Pigeons)

These two carrier pigeons were awarded the Dickin Medal in February 1947 for their service to the Australian Armed Forces during World War II.

Number DD.43.T.139, a blue bar cock, saved 1402 Australian troops and their ammunition on July 12, 1945, when their ship, *Army Boat*, crashed into the coast of Wadou Beach in the Solomon Sea. Being isolated, their only option was to send this pigeon with a message requesting a rescue boat. The pigeon flew 40 miles in 50 minutes through the torrential rains and gusting winds of a tropical storm, successfully delivering the message to Madang, and the soldiers were saved. In total this bird made 23 successful flights covering a distance of 1,002 miles.

Number DD.43.Q.879 served with the Australian Signal Corps during World War II, and as a result of this pigeon's heroic flight, U.S. forces were able to successfully defend themselves from a Japanese attack. On April 5, 1944, three pigeons were released with messages to warn headquarters of an eminent Japanese attack on a U.S. Marine patrol on Manus Island. Flying through heavy gunfire DD.43 was the only one of the three to make it alive to headquarters.

Tyke (Elephant)

Tyke, an African bush elephant from Mozambique, was performing with Circus International of Honolulu, Hawaii, in 1994. On August 20, 1994, a terrible series of events took place, ultimately leading to Tyke's death. During a performance at the Neal Blaisdell Center, Tyke went on a rage, killing her trainer, Allen Campbell, and seriously injuring her groomer, Dallas Beckwith; she also attacked publicist Steve Hirano while he was trying to prevent her from fleeing the parking lot. Afterwards she burst out of the arena and ran for more than thirty minutes through the streets of the Kakaako business district until local police fired 86 shots at the elephant and she collapsed and died. Tyke had had a background of three other incidents in the past. On April 21, 1993, during a performance, Tyke broke free through the doors of the Altoona, Pennsylvania, Jaffa Shrine Center in a rampage, causing $14,000 in damages. The day after, on April 22, a circus worker reported that the elephant had attacked a tiger trainer. On July 23, 1993, at the North Dakota State Fair in Minot, Tyke trampled and injured a handler and took off running for twenty-five minutes. USDA and Canadian law enforcement reported that while an elephant named Tyke (possibly the same elephant) was performing

with the Tarzan Zerbini Circus, the handler was publicly observed beating the elephant, causing it to scream and bend down to avoid being hit. Afterwards when the handler walked by, the elephant screamed and veered away. In the aftermath of the August 20 incident, Tyke became a symbol for animal rights and inspired a campaign to end animal abuse in the entertainment industry. A footnote: Campbell's autopsy reported that cocaine and alcohol were in his system, and also officials at the Denver Zoo reported that in the 1980s there had been complaints of animal abuse against Campbell while he was operating elephant and camel rides.

Tyke (Pigeon)

Tyke, also known as "George," service number 1263 MEPS 43, was a male homing pigeon during World War II. He was hatched from British and South African parents in Cairo, Egypt, and served with the Middle East Pigeon Service. In June 1943, Tyke was serving on an American bomber and was released with a message to get help after the plane was shot down. He flew over a hundred miles and successfully delivered the message for help to friendly forces. The bomber crew gave Tyke credit for saving their lives. In 1943, Tyke was awarded the Dickin Medal for gallantry. His citation reads, "For delivering a message under exceptionally difficult conditions and so contributing to the rescue of an Air Crew, while serving with the RAF in the Mediterranean in June 1943." Tyke was one of the first pigeons, along with White Vision and Winkie, to be awarded the Dickin Medal. In July of 2000, Tyke's Dickin Medal was auctioned by Spink in London and sold for $7,313.

Unsinkable Sam (Cat)

With the original name of Osker or Oscar, Sam was born before 1941 in

Germany and died in 1955 in Northern Ireland. Sam's story begins on the German battleship *Bismarck*. In May 1941, the *Bismarck* was on an operation to raid routes from North America to Great Britain when it was spotted by Allied Forces and attacked in the Battle of Denmark Strait. The *Bismarck* destroyed the British battlecruiser HMS *Hood*, and on May 27, 1941, the British battleship HMS *Cossack* sunk the *Bismarck*, killing 2,200 German soldiers; only 115 people and one black and white cat survived. The British vessel found Unsinkable Sam floating on a board. While aboard, Sam was a moral support to the crew of the HMS *Cossack* until October 24, 1941, when a German submarine fired a torpedo, severely damaging the ship and killing 159 sailors. True to his name, Unsinkable Sam survived and subsequently was transferred to the HMS *Ark Royal* aircraft carrier. Once again Sam survived an attack by a German submarine, U-81, that brought down the HMS *Ark Royal* 30 miles from Gibraltar on November 14, 1941. A motor launch found Sam clinging to a floating plank. After the loss of the *Ark Royal*, Sam was sent back to the United Kingdom. Unsinkable Sam survived the destruction of three major vessels during World War II and died in 1955. Some historians question whether one cat could survive all three shipwrecks.

Upstart (Horse)

Upstart, a chestnut gelding, was one of 186 horses that were part of the Metropolitan Police Mounted Branch in London during World War II. These horses were used to aid civilians during the Blitz and for much of the war, from September 1940 to late 1944. Upstart was stabled near Hyde Park when a nearby bomb attack damaged his stable. Upstart and his handler, DI J. Morley, relocated to East London and were patrolling a street in Bethnal Green when a bomb landed 75 feet in front of Upstart,

showering them with glass and shrapnel. Upstart helped Morley with directing traffic and crowd control. On April 11, 1947, Upstart was awarded the Dickin Medal in a ceremony at Hyde Park.

Ushuaia *see* Herr Olsson

Verdun Belle (Dog)

This setter wandered into an American regiment in France and attached herself to a young marine who bathed her and shared his dinner with her. Under a coat of mud was a silky haired white dog with brown patches. The marine named her Verdun Belle (Verdun is the town that was occupied by the regiment). She slept at his feet and accompanied him on duty. He modified a gas mask for her and soon she learned to fetch it before a gas attack. The following spring she gave birth to nine puppies. The marine took the puppies with him on his orders to advance across France. However, taking all nine puppies proved too difficult and all but two of the puppies died. In the confusion Verdun Belle was lost. The marine headed into battle and left the two puppies with an ambulance passing by. Verdun Belle turned up following another regiment of Marines. She was reunited with her surviving pups at a nearby farm, where a field hospital had been set up. One evening a shell-shocked marine arrived unconscious and the dog went crazy licking his face. The reunion was complete.

Virginia (Horse)

Major General Jeb Stuart believed that his horse, Virginia, saved him from being captured by jumping over a large ditch during the Civil War Battle of Hanover.

Von Kluck (Dog)

Von Kluck was one of the enemy's dogs that were captured and retrained to become great messenger dogs. It is reported that on one mission Von Kluck was bombed and thrown into the air and amazingly got up and finished his mission to the surprise of the soldiers.

Vonolel (Horse)

Vonolel was purchased in India in 1877 by British Commander Frederick Roberts. During his military career this white Arab charger horse was decorated with three medals including the Kandahar Star. The medal was presented commemorating the Second Anglo-Afghan War in 1880. Vonolel died in 1899 at age 29 at the Royal Hospital in Kilmainham and was buried with full military honors. The poem on his headstone honors his military service as well as his character. It reads: "Dumb creatures we have cherished here below/shall give us joyous greeting when/ We pass the golden gate/Is it folly that I hope it may be so."

Voytek *see* Wojtek

Waghya (Dog)

Waghya (the name means "tiger" in Marathi) was a mixed breed and loyal pet of Maratha emperor Chhatrapati Shivaji Maharaj. The dog was "known as the epitome of loyalty and eternal devotion." It is said that after the emperor's death in 1680, the mourning dog jumped into the funeral pyre and sacrificed himself. There is a statue of Waghya on a pedestal next the Shivaji's tomb at Raigad Fort.

Wallace (Dog)

Wallace is known as "The Canine Who Wanted to Be a Fire Dog." After Wallace passed away in 1902, many mourned him as the dog "who worked his wee paws off for seven years, leading the fire Brigade back and forth from the fires in the city, day and night." Wallace more or less joined the Fire Brigade when he followed the brigade at a fundraising parade for the Lifeboat Funding Appeal through the streets of Glasgow in 1894. Even though his owner kept bringing him back home, Wallace kept returning to the Central Fire Station. Eventually he ended up staying and he spent the rest of his life as their mascot and helper. Wallace would go with the men to fires, running ahead to warn people to get out of the way of the coming wagon. Spectators were puzzled as to how Wallace always knew his way to the location of the fires. In reality he took his directions from the driver, who used his whip (not on Wallace) to signal right or left turns. People often visited Wallace at the station, and one elderly woman bought four rubber boots, made especially for him, to protect his sore paws, but unfortunately he lost two boots the first time he wore them on a call. After his death the firemen had him embalmed and placed in a glass case with his two rubber boots and wearing a collar with officer's rank markings; he is on display at the Scottish Fire and Rescue Service Museum & Heritage Centre in Greenock. Poems were written as tribute to him and several books have been written about him.

Warrior (Horse)

In 2014, Warrior, "the horse the Germans could not kill," was awarded posthumously the Honorary PDSA Dickin Medal on the 100th anniversary of World War I, on behalf of all animals who served the British forces during the war. The Dickin Medal is the equivalent of the Victoria Cross awarded to people. Warrior served faithfully with his inseparable owner, Major General Jack Seely, on the Western Front from 1914 to 1918. He faced entrapment in burning stables, air attacks and shellings and took part in the battles of the Somme and Passchendaele. In March 1918,

Seely and Warrior, along with the Canadian Cavalry Brigade, took part in the last major battle of the war, at Moreuil Wood. Warrior lived out his life with the Seely family at the Isle of Wight until his death at age 33. There is a painting of Warrior and Seely by artist Gilbert Joseph Holiday.

Washoe (Chimpanzee)

Captured as an infant in Africa by the U.S. Air Force for the purpose of research, Washoe was about a year old when behavioral psychologists R. Allen and Beatrice Gardner at the University of Nevada at Reno acquired her in June 1966. The Gardners named her Washoe for their Nevada county and began raising her in a trailer. Washoe became the first non-human to learn to communicate in American Sign Language. In 1971, primate researcher Dr. Roger Fouts moved Washoe to the Institute for Primate Studies in Norman, Oklahoma. For the first time, she met other chimpanzees. In 1979, Washoe adopted an 11-month-old male chimpanzee named Louis and began teaching him signs. Washoe died at the age of 42 in 2007.

Wee Jock (Dog)

Wee Jock, an Irish terrier, was awarded the Purple Cross posthumously in a ceremony on November 30, 1997. Wee Jock's owner, nicknamed "The Pikeman," was an Irish miner fighting against the Crown forces in the Eureka Rebellion in Ballarat, Victoria, in 1854. While guarding the Irish flag, The Pikeman was killed; Wee Jock had showed total devotion by staying by his side for the entire battle. Reportedly, Wee Jock even stayed with The Pikeman's body on the death cart while it was being transported to the cemetery. In addition to the award, Wee Jock is also honored as a bronze centerpiece of the statue at Eureka Stockade Memorial Park.

Wheely Willy (Dog)

Wheely Willy's message, "Life is what you make of it," will never be forgotten. This remarkable little Chihuahua was born in 1991 in Long Beach, California. When he was very young, his owner abused and even attempted to murder him. Found in a box, he had incurred spinal injuries and a cut throat and was saved at a veterinary hospital, but would never walk again. Wheely Willy was adopted by groomer Deborah Turner, who did everything to give him a happy life despite his difficulties. When K-9 carts became available, Willy adapted quickly and was able to finally move around. The local media made him a national celebrity and he was featured on the Animal Planet network and several talk shows. Wheely Willy also made public appearances to promote understanding for people with physical disabilities, and he traveled the world visiting hospitals, lifting the spirits of children. He was very popular in Japan and caused a sensation when Prince Hitachi and Princess Hanako got down on the floor upon meeting him. Wheely Willy died on December 22, 2009, from injuries he sustained when his owner slipped on her porch and he fell out of her arms. Wheely Willy is the subject of two bestselling children's books.

White Vision (Pigeon)

This female World War II homing pigeon served with the National Pigeon Service and was posted to No. 190 Squadron RAF. She was awarded the first Dickin Medal for gallantry in 1943 for delivering a message from a flying boat to a ditch off the coast of Scotland. With the plane's radio out of order, she was released to indicate where the plane had gone down. She flew 60 miles to deliver the message against a strong headwind. She arrived exhausted with many missing feathers. Her Dickin Medal citation reads, "For delivering a message under exceptionally difficult conditions and so contributing to the rescue of an Air Crew while serving with the RAF in October 1943." White Vision died in 1953.

Whitefoot (Dog)

Handler Errington tells of how his three dogs, Jack the Airedale, and two large Welsh terriers, Whitefoot and Lloyd, worked in Strazeele (in northern France) through difficult roads littered with the remnants of shelling.

William "Billy" Windsor (Goat)

Queen Elizabeth liked Billy, recommending that he become a member of the Royal Welsh 1st Battalion. Billy was a salaried lance corporal with Army number 25232301. Billy received daily rations of Guinness and two cigarettes. After eight years of service, he retired with honors. The one stain on his impeccable record was when he headbutted a drummer at the queen's birthday parade, for which he was temporarily demoted to private.

William Johnson see Billy

William Johnson Hippopotamus see Billy

Willie (Parrot)

Willie, a green-feathered Quaker parrot, belonged to two-year-old Hannah's babysitter in Denver. One day in 2008, Hannah was eating while her babysitter was in the bathroom. Suddenly Hannah began choking and Willie started to shout "Mama, Baby, Mama," flapping his wings. The babysitter rushed in and performed the Heimlich maneuver in time. Willie was presented with the Animal Lifesaver's Award by the Red Cross at a ceremony attended by both the governor of Colorado and the mayor of Denver.

Winchester see Rienzi

Winkie (Pigeon)

During World War II, Winkie served aboard a British Bristol Beaufort bomber. Following a mission to Norway on February 23, 1942, a badly damaged RAF bomber crashed into the North Sea 100 miles from home. The crew of four men were in freezing water, and, not being able to radio their position, they looked to their blue checkered hen, number NEHU 40 NSL, named Winkie, to fly home. Winkie flew 120 miles to Broughty Ferry to her owner George Ross, who assembled a search and rescue mission. Calculating the time difference between the crash and the arrival of the pigeon, and also the wind speed and the inhibition of her flight speed caused by oil on her feathers, it was estimated a rescue ship had been dispatched within 15 minutes. On December 2, 1943, Winkie and two other pigeons were the first recipients of the Dickin Medal. Winkie's citation reads, "For delivering a message under exceptional difficulties and so contributing to the rescue of an Air Crew while serving with the RAF in February 1942."

Winnie see Winnipeg

Winnie (Cat)

One night in 2007, fourteen-year-old Winnie, a golden-eyed tiger cat, saved her family from carbon monoxide poisoning in New Castle, Indiana. At about 1:00 a.m. Winnie jumped on Eric and Cathy Keesling's bed, meowing wildly and nudging them. Cathy woke up in a daze and when she couldn't wake her husband she called 911. Paramedics arrived, saving them and their unconscious son, Michael. It seems that a gasoline-powered water pump in the basement had caused the odorless gas to build up. The ASPCA named Winnie the "Cat of the Year." This wasn't the first time

Winnie saved her family's life; she acted similarly when tornadoes were in the area.

Winnipeg (Bear)

Winnipeg "Winnie" was an American black bear born in 1914 in Ontario, Canada. She is mostly recognized as the inspiration for A.A. Milne's Winnie the Pooh character. She was purchased as a cub by Captain Harry Colebourn for $20 at a train stop in White River, Ontario, while he was en route to Valcartier to volunteer with the Canadian Army Veterinary Corps (CAVC). The bear's mother had been killed by a trapper. Colebourn named her "Winnie" after his hometown of Winnipeg, Manitoba. The hunter who sold her is thought to be responsible for the bear's early socialization and is not documented. Winnie went on to be quite a busy bear, becoming a mascot to the CAVC and also a pet to the Second Canadian Infantry Brigade Headquarters. When Harry was sent to the war in France, he felt it would be too dangerous for the little bear cub, so he donated her to the London Zoo. Author Lindsay Mattick, Harry's great-granddaughter, recounted the bear's journey through an army base in England to the London Zoo, where Winnie met A.A. Milne's son Christopher Robin. Winnie died at the London Zoo on May 12, 1934, at the age of 20. Her skull was kept in the Odontological Collection at the Royal College of Surgeons Hunterian Museum in London. Her skull went on public display in 2015. Several statues and plaques honoring her life can be found in London and Winnipeg. The London Zoo has a statue of Winnie by sculptor Lorne McKean. A life-size statue of Harry Colebourn and Winnie by sculptor William Epp stands at the Assiniboine Park Zoo in Winnipeg. In 1996, Canada Post issued a 45-cent stamp, "Winnie and Lieutenant Colebourn, White River, 1914," designed by Wai Poon, with art direction by Anthony Van Bruggen, computer

Winnie with her owner Captain Harry Colebourn in 1914. Winnie was the inspiration for A.A. Milne's "Winnie the Pooh" character (Manitoba Provincial Archives).

design by Marcelo Caetano and printed by Ashton-Potter Canada Limited. Winnie's story was portrayed in the 2004 movie *A Bear Named Winnie* starring Michael Fassbender as Colebourn. Bonkers, a male American black bear, played an adult Winnie. In 2011, author M.A. Appleby wrote a children's book titled *Winnie the Bear*. Another children's book, *Finding Winnie: The True Story of the World's Most Famous Bear*, written by Lindsay Mattick and illustrated by Sophie Blackall, was published by Little, Brown and Company.

Wojtek (Bear)

The name "Wojtek" is a diminutive form of an old Slavik name, "Wojciech," meaning "he who enjoys war" or "joyful warrior." Wojtek, a Syrian brown bear, began his army career when a young Iranian boy found him as a cub, and he was purchased by Lieutenant Anatol Tarnowiecki in April 1942. After spending three months in a Polish refugee camp, Wojtek was donated to the 22nd Artillery Supply Company. Wojtek was drafted into the Polish Army as a private, primarily for his rations and transport. There are numerous accounts that he helped to move crates of ammunition during the Battle of Monte Cassino in Italy in 1944. After the war he was stationed with the 22nd Company at the Winfield Airfield on Sunwick Farm near an area called The Scottish Borders. On November 15, 1947, he was given to the Edinburgh Zoo, where he spent the rest of his life. Beer was his favorite drink and he also smoked (or ate) cigarettes. Wojtek died at age 21 on December 2, 1963. He is memorialized with plaques at the Imperial War Museum in London; a sculpture by David Harding at the Sikorski Museum in London; a carved wooden sculpture in Weelsby Woods, Grimsby; a statue in Park Jordana, Krakow; and a bronze statue by Alan Beattie Herriot of Wojtek walking with a Polish soldier, in Princes Street Gardens, Edinburgh. Other memorials include a film, *Wojtek-The Bear That Went to War*, broadcast on BBC on December 30, 2011; a book by Aileen Orr, *Wojtek: The Bear*, with an epilogue by Neal Ascherson and published by Birlinn Limited in 2014; and a music video, "Wojtek," by British singer and songwriter Katy Carr. Wojtek is referenced in the computer game *Hearts of Iron IV* and in one of the card artworks in the Scythe board game. According to The Soldier Bear website, Wojtek "became part of the history of the Polish Armed Forces in the Second World War and his legacy will endure."

Wolf (Dog)

Wolf (Brand Number T121) proved to be an invaluable member to the 27th Infantry in their battle through the Corabello Mountains in Northern Luzon to the strategic position on Belete Pass. The patrol proceeded with great caution, for the rough terrain of the jungle could easily hide a Japanese regiment. While on patrol, Wolf could pick up the enemy's scent as far as 150 yards ahead. While proceeding forward, Wolf suddenly stopped and pointed toward a hill. The soldiers opened fire and the Japanese fired back from the hill Wolf had indicated. At one point the patrol was encircled by the enemy and, taking Wolf's lead, the patrol safely made their way down the mountain back to the command post. During a fire fight Wolf sustained several shrapnel wounds, which he managed to hide from the men. Despite excellent medical care and an emergency operation, Wolf, U.S. Army War Dog T121, died from his wounds.

Xiang Xiang (Giant Panda)

Successful breeding programs with pandas in captivity has contributed to the jump in population over the past 30 years, while pandas in the wild continue to be in crisis. Panda experts say there needs to be more research on the lifecycle of pandas in the wild, since, despite no biological differences between pandas raised in captivity and those in the wild, they cannot seem to live together. Xiang Xiang was born in captivity on August 25, 2001, at the Wolong Giant Panda Research Center in Sichuan Province. In April 2006, he became the first giant panda bred and raised in captivity to be released into the wild. Xiang Xiang wore a radio collar so that researchers could monitor his feeding habits and movements. Despite three years of training to equip him with the necessary skills to survive in the wild, on February 19, 2007, he was found dead—less than a year after his release. It was speculated that he fell from a tree or died running from other pandas; Zhang Hemin, chief of the Wolong center, surmised that "Xiang Xiang died from fighting for territory or over a female mate with other male pandas." Tao Tao, a male panda born on August 3, 2010, in a semi-wild training base in Hetaoping, may be the next panda to be released into the wild. However, before he can be released, he must excel in three different sized training fields.

Yank (Pigeon)

Serving during World War II, Yank flew a 98-mile mission to the headquarters at Tebessa, Algeria, to deliver news that American soldiers had retaken Gafsa, Tunisia.

Yew-Yew (Dog)

Yew-Yew was a white shaggy mutt in service for two years, leading field nurses to hundreds of wounded soldiers. During her service, she was wounded, losing an ear and an eye.

Yorick (Monkey)

Yorick, previously known as Albert VI, survived his 45-mile flight in 1951. However, while waiting to be released from his metal capsule, he died from heat exhaustion under the hot New Mexico sun.

York (Dog)

York was a setter pet of Brigadier General Alexander S. Asboth, who often took him into action.

York (Dog)

The story of York (Brand Number 011X) does not end with his service as a scout dog with the 26th Infantry Scout Dog Platoon in Korea, but also continues upon his return to the United States. He did have a brilliant career in Korea, taking part in 148 combat patrols between June 12, 1951, and June 26, 1953. York was even given a Distinguished Service Award by General Samuel T. Williams. After returning to the United States, he was used as part of a demonstration team at the Army Dog Training Center at Fort Carson, Colorado, to improve public support for recruitment and procurement of dogs to be used in the military. After the training center at Fort Carson was closed on July 1, 1957, York was transferred to the 26th Infantry Scout Dog Platoon at Fort Benning, Georgia.

Yorkie Doodle Dandy see Smokey

You'll Do Lobelia see Elsie

Yuki (Dog)

Yuki (from the Japanese word for "snow") was a stray terrier mix found at a gas station in Texas in 1966 by President Lyndon B. Johnson's daughter Luci on her way for Thanksgiving dinner at his ranch. Yuki became President Johnson's favorite dog; the president brought him to the Oval Office and to Cabinet meetings. The two had such a strong bond that they swam together in the pool, and Yuki was President Johnson's dance partner at his daughter Lynda's wedding. The two would show off their singing talent by lifting their noses in the air and howling together. After Johnson left office in 1969, Yuki went to live at the family's ranch until Johnson passed away, and he then went to live with Luci's family until he died in 1979.

Za (Dog)

Za, an Alsatian sheepdog, served at the front lines during World War I. Along with Helda, another Alsatian sheepdog, Za located a German outpost under the guidance of Sergeant Megnin and his assistant.

Zanjeer (Dog)

Zanjeer, a Labrador retriever born on January 7, 1992, whose name comes from the 1973 Hindi film *Zanjeer*, was also called "Ginger" after his coat color.

Zanjeer was trained to be a detection dog at the Dog Training Centre of the Criminal Investigation Department at Shivaji Nagar in Pune, India. Afterward, on December 29, 1992, he joined the Mumbai Police Bomb Detection and Disposal Squad being handled by Ganesh Andale and VG Rajput. Starting on March 15, 1993, Zanjeer proved to be indispensable in preventing at least three attacks during the 1993 Mumbai bombings. Overall, during his service, Zanjeer recovered 11 military bombs, 57 country-made bombs, 175 petrol bombs, and 600 detonators. Zanjeer died from bone cancer on November 16, 2000, and was buried with full state honors.

Zarafa (Giraffe)

Zarafa, a female Nubian giraffe, was a diplomatic gift from Muhammad Ali of Egypt. Two years after leaving Egypt, Zarafa was finally presented to King Charles X of France on July 9, 1827. In those two years she had quite a journey. Beginning in southern Sudan, she traveled on a custom-designed barge 2,000 miles northward along the Nile to Alexandria. She was accompanied by three cows, providing her with 25 liters of milk each day. From Alexandria, after a month-long sea voyage, Zarafa arrived in Marseille on October 31, 1826. Fearing that it might be too dangerous to transport her to Paris by boat, it was decided that she would walk the 900 kilometers to Paris. After spending the winter in Marseille, she set off on her journey on May 20, 1827, accompanied by her cows and Étienne Geoffroy Saint-Hilaire (scientist and head of the zoo at Jardin des Plantes). Forty-one days later, Zarafa was presented to the king and made her home in the Jardin des Plantes. Being the first live giraffe the people of France had ever seen, Zarafa drew massive curious crowds at the Jardin des Plante Zoo. After her arrival, everything "giraffe" became popular—books were written about her and paintings were made; spotted fabric in the color of "belly of giraffe" was all the rage; and giraffe images were painted on porcelain and ceramics. Zarafa remained in Paris for 18 years until her death on January 12, 1845. Her stuffed corpse remains at the Museum of Natural History of La Rochelle. American author Michael Allin named her Zarafa in his 1998 book *Zarafa: A Giraffe's True Story, From Deep in Africa to the Heart of Paris*. In Arabic *zarafa* means "charming" or "lovely one" and is a variation of the Arabic *zerafa*. During her journeys, Zarafa was dubbed *le bel animal du roi* ("The Beautiful Animal of the King") and *la Belle Africaine* ("the Beautiful African"). In 1827, Muhammed Ali gifted two other giraffes: one to George IV of the United Kingdom that survived less than two years; and the other one to Francis I of Austria that survived less than one year. Purple House Press published a children's book inspired by Zarafa's story, *The Giraffe that Walked to Paris*, written by Nancy Milton and illustrated by Roger Roth.

Zarif *see* **Muhamed**

Bibliography

Abadzis, Nick. *Laika*. New York: Square Fish, 2014. Graphic novel of fictional account of Laika.

Adamson, George. *My Pride and Joy*. New York: Simon & Schuster, 1987.

Adamson, Joy. *Born Free*. New York: Pantheon Books, 1960.

Ahlbrandt, Arlene Briggs. *Annie the Railroad Dog, A True Story*. Fort Collins, CO: Citizen Printing, 1998.

Alberti, Samuel. *The Afterlives of Animals: A Museum of Menagerie*. Charlottesville: University of Virginia Press, 2013. This collection of articles contains biographies of animals, especially their posthumous roles as museum specimens.

Alexander, Caroline. *The Endurance: Shackleton's Legendary Antarctic Expedition*. New York: Knopf, 1998.

Alexander, Caroline. *Mrs. Chippy's Last Expedition: The Remarkable Journal of Shackleton's Polar-Bound Cat*. New York: Harper Perennial, 1999.

Allin, Micheal. *Zarafa: A Giraffe's True Story from Deep in Africa to the Heart of Paris*. London: Headline, 1999. Tale of Zarafa travels from Africa to Paris.

Ambrose, Stephen. *Nothing Like It in the World: The Men Who Built the Transcontinental Railroad, 1863–1869*. New York: Simon & Schuster, 2001.

Ambrose, Stephen. *Undaunted Courage: Meriwether Lewis, Thomas Jefferson, and the Opening of the American West*. New York: Simon & Schuster, 1996.

Amsden, Roger. "Chinook Disappears on Byrd's Antarctic Expedition on his 12th Birthday." *The Laconia Daily Sun*, December 28, 2016.

"Animal Attraction: Mascots Brightened the Dark Corners of War." *World War II*, July-Aug. 2013, p. 48+. *General OneFile*, http://link.galegroup.com/apps/doc/A332788239/GPS?u=vol_e7&sid=GPS&xid=ef7b7205.

"Animals in the Great War." *MHQ: The Quarterly Journal of Military History*, Autumn 2017, p. 94. Rare photographs of animals during World War I.

"Animals Within the War." *Belfast Telegraph* [Belfast, Northern Ireland], 5 Nov. 2014, p. 22. *Infotrac Newsstand*, http://link.galegroup.com/apps/doc/A388685667/GPS?u=vol_e7&sid=GPS&xid=f1a27236.

Armistead, Gene. *Horses and Mules in the Civil War: A Complete History with a Roster of More Than 700 War Horses*. Jefferson, NC: McFarland, 2013. Information about 700 horses/mules during the Civil War.

Armstrong, Jennifer. *Shipwreck at the Bottom of the World*. New York: Knopf Books for Young Readers, 2000.

Aronson, Deb. *Alexandra the Great: The Story of the Record-Breaking Filly Who Ruled the Racetrack*. Chicago Review Press, 2017.

Aubrey-Herzog, Jay. "Battle for Boomer Jack." *North Coast Journal*, May 31, 2007.

Barloy, Jean Jacques. *Man and Animal*. London: Gordon & Cremonesi, 1978. 100 centuries of man's relationship with animals, cooperation, domestication and friendship.

Barnes, Michael. "The Short Suborbital Career of Austin Native Sam the Space Monkey." *Lifestyle*, January 21, 2018.

Bausum, Ann. *Sergeant Stubby: How a Stray Dog and His Best Friend Helped Win World War I and Stole the Heart of a Nation*. Washington, D.C.: National Geographic, 2015.

Baynes, Ernest Harold. *Animal Heroes of the Great War*. New York: Macmillan Co., 1933. Moving stories of animals during World War I.

Bedini, Silvio. *The Pope's Elephant*. New York: Penguin Books, 1997.

Belozerskaya, Marina. *The Medici Giraffe: And Other Tales of Exotic Animals and Power*. New York: Little, Brown and Co., 2009. History of influential role of exotic animals as diplomacy gifts and treasures.

Blakeslee, Nate. *American Wolf: A True Story of Survival and Obsession in the West*. New York: Crown Publishing, 2017.

Bourke, Anthony, and John Rendall. *A Lion Called Christian*. New York: Broadway Books, 1971.

Bowan, Bob. *A Street Cat Named Bob*. London, England : Hodder and Stoughton, 2012.

Bowan, Bob. *The World According to Bob*. London, England : Hodder and Stoughton 2013.

Bronstein, Phil. "The Shooter." *Esquire*, March 2013.

Brouwer, Sigmund. *Innocent Heroes: Stories of Animals in the First World War*. Plattsburgh: Tundra Books, 2017. Fictional stories based on some truth.

Bulanda, Susan. *Soldiers in Fur and Feathers: The Animals That Served in World War I*. Crawford, CO: Alpine Press, 2014. How the Western Allies used animals.

Campbell, Clara. *Bonzo's War: Animals Under Fire 1939–1945*. London: Constable, 2013. Fate of pets during World War II.

Campbell, Clare. *Dogs of Courage: When Britain's Pets Went to War 1939–45*. London: Corsair, 2016. Stories about pets during World War II in Britain.

Carey, Benedict. "Alex, a Parrot Who Had a Way with Words, Dies." *The New York Times*, September 10, 2007.

Chaline, Eric. *Fifty Animals That Changed the Course of History*. Buffalo: Firefly Books, 2011. Guide to animals that had the greatest influence on human civilization.

Chambers, Paul. *Jumbo: The Greatest Elephant in the World*. Hanover, NH: Steerforth Press, 2008. Incredible tale of Jumbo the elephant.

Chambers, Paul. *A Sheltered Life: The Unexpected History of the Giant Tortoise*. New York: Oxford University Press, 2005. Compelling history of the giant tortoise and Lonesome George.

Clark, Eddie. "Twenty-five Years with Andre," *Yankee Magazine*, November 1986.

Cooper, Gwen. *Homer's Odyssey: A Fearless Feline Tale, or How I Learned About Love and Life with a Blind Wonder Cat*. New York: Delacorte Press, 2009.

Cooper, Jilly. *Animals in War*. London: Corgi, 2010. Stories of brave animals that played a role in wars.

Coren, Stanley. *The Pawprints of History: Dogs and the Course of Human Events*. New York: Free Press, 2002.

Croke, Vicki. *Elephant Company: The Inspiring Story of an Unlikely Hero and the Animals Who Helped Save Lives in World War II*. New York: Random House, 2014.

Croke, Vicki Constantine. *The Lady and the Panda*. London: Nicholson and Watson, 2005. The true story of the first American explorer to bring back China's most exotic animal.

DeHovitz, Ross E. "The 1901 St. Louis Incident: The First Modern Medical Disaster," *Pediatrics*, 2014.

"Dog Star: Scientist Recalls Training Laika for Space." *Bangladesh Government News*, November 3, 2017. *Infotrac Newsstand*, http://link.galegroup.com/apps/doc/A513130039/GPS?u=vol_e7&sid=GPS&xid=d12c5dff.

Duncan, Dayton, and Ken Burns. *Lewis & Clark: An Illustrated History*. New York: Alfred Knopf, 1997. Based on the PBS documentary film by Ken Burns.

Durrell, Gerald Malcolm. *The Overloaded Ark*. London: Hart-Davis, 1953.

Eglan, Jared. *Beasts of War: The Militarization of Animals*. San Bernardino, CA: Lulu.com, 2015. Collection of articles from Wikipedia about use of animals during war.

Elmore, Ronnie G. "Candidates' Pets Can Make a Difference at the Ballot Box." *Veterinary Medicine*. November 2008.

Fagan, Brian M. *The Intimate Bond: How Animals Shaped Human History*. New York: Bloomsbury Press, 2015. Shows how the powerful bond between humans and animals has shaped civilization.

Flinders, Matthew. *Trim: The Story of a Brave, Seafaring Cat*. Australia: HarperCollins Publishers, Aug. 1, 1997.

Foley, George F., and George Gray. *Sinbad of the Coast Guard: The True Story of the Ultimate Sea Dog*. Mystic, CT: Flat Hammock, 2005.

Frankel, Rebecca. "The Dog Whisperer: How a British Colonel Altered the Battlefields of World War I, and Why His Crusade Still Resonates Today." *Foreign Policy*, Sept.-Oct. 2014, p. 90+. *General OneFile*, http://link.galegroup.com/apps/doc/A382808190/GPS?u=vol_e7&sid=GPS&xid=105104b8.

Frankel, Rebecca. "Dogs in War." *National Geographic*, May 18, 2014.

Frankel, Rebecca. *War Dogs: Tales of Canine Heroism, History, and Love*. New York: St. Martin's Press, 2015.

Galvani, William. "Sea Dogs." *American Heritage*, October 1994.

Gardiner, Juliet. *The Animals' War: Animals in Wartime from the First World War to the Present Day*. Published in association with the Imperial War Museum to coincide with their major exhibition.

Gardiner, Juliet. "Rossetti's Wombat: Pre-Raphaelites and Australian Animals in Victorian London." *History Today*, July 2009.

George, Isabel. *Beyond the Call of Duty: Heart-Warming Stories of Canine Devotion and Wartime Bravery*. London: HarperElement, 2010. Dogs that made a difference during wartime.

George, Isabel. *The Dog That Saved My Life*. London: HarperElement, 2010. Features tales of canine bravery in wartime.

George, Isabel. *Murphy the Hero Donkey: A True WWI Story*. London: Harper, Collins Publisher, 2015.

Gill, Gillian. *Nightingales: The Extraordinary Upbringing and Curious Life of Miss Florence Nightingale*. New York: Random House, 2004.

Goessling, Michael. "A Very Big Deal: Thumbelina's Calling." *The Exceptional Parent*, March 2009.

Goodavage, Maria. *Soldier Dogs: The Untold Story of America's Canine Heroes*. New York: New American Library, 2012. Sampling of military dogs and their achievements.

Greenhill, Sam. "Still Sprightly at 183, the Oldest Creature on Earth." *Daily Mail* (London, England), January 8, 2016.

Greenwood, Mark. *Midnight: A True Story of Loyalty in World War 1*. Australia: Walker Books, 2014.

Grier, Katherine. *Pets in America: A History*. Chapel Hill, NC: University of North Carolina Press, 2006. Entertaining history of pets in America.

Grigson, Caroline. *Menagerie: The History of Exotic Animals in England, 1100–1837*. New York: Oxford University Press, 2016. Detailed and entertaining tales of exotic animals in Europe.

Hall, Karen. "Civil War Rooster's Portrait to Be Displayed." *Marshall County Tribune*. Aug. 5, 2011.

Harris, Paul. "Bing, the Dog of War Who Parachuted into France to Become a D-Day Hero," *Daily Mail*, April 17, 2012.

Hatkoff, Isabelle. *Owen & Mzee: The True Story of a Remarkable Friendship*. New York: Scholastic Press, 2006.

Hayter-Menzies, Grant. *From Stray Dog to World War I Hero: The Paris Terrier Who Joined the First Division*. Washington, D.C.: Potomac Books, 2015.

Heatherly, Chris. "Reckless: The Racehorse Who Became a Marine Corps Hero." *Military Review*, January-February 2017.

Herman, Carol. "The Elephant They Loved." *The Washington Times* (Washington, D.C.), March 9, 2008.

Hillenbrand, Laura. *Seabiscuit: An American Legend*. New York: Ballantine Books, 2002.

Houser, Sue. *Hot Foot Teddy, The True Story of Smokey Bear*. Evansville, IN: M.T. Publishing Company, Inc., 2007. A portion of the sales are donated to the Smokey Bear Forest Fire Prevention program.

Hutton, Robin L. *Sgt. Reckless: America's War Horse*. Washington, D.C.: Regnery, 2014. Incredible tale of Sgt Reckless in the Korean war.

Janus, Allan, *Animals Aloft*, Piermont, New Hampshire: Bunker Hill Publishing, 2005.

Jedick, Peter. "Smoky the War Dog: A Tiny Terrier Found in a New Guinea Foxhole Became a Pint-Sized War Hero Because of a GI's Friendship." *America in WWII*, August 2008, p. 52+. *General OneFile,* http://link.galegroup.com/apps/doc/A401095993/GPS?u=vol_e7&sid=GPS&xid=a6a56317

Jensen, Karen. "For the Birds: A Thousands-Strong Force of Messengers, Spies, and Counterspies Helped Save Lives and Win Battles Throughout the War." *World War II*, December 2008.

Keeney, Douglas. *Buddies: Men, Dogs, and World War II*. Osceola, WI: MBI Pub. Co., 2001. Photographs of servicemen with their dog mascots.

Kehe, Marjorie. "The Odd, Remarkable, Yet True Story of Jumbo the Circus Elephant." *The Christian Science Monitor,* April 1, 2008.

Kiser, Toni, and Lindsey F. Barnes. *Loyal Forces: The American Animals of World War II*. Baton Rouge, LA: Louisiana State University Press, 2013. Details use of American animals during World War II.

Klein, Christopher. "The Thanksgiving Raccoon That Became a Presidential Pet." *History Stories,* November 18, 2016.

LaLande, Jeffrey M. *Prehistory and History of the Rogue River National Forest: A Cultural Resource Overview*. Medford, OR: USDA Forest Service, 1989.

Le Chene, Evelyn. *Silent Heroes: The Bravery and Devotion of Animals in War*. London: Souvenir, 2009. Profile of animals.

Lemish, Michael. *War Dogs: A History of Loyalty and Heroism*. Washington, D.C.: Potomac, 2008.

Leslie-Melville, Betty and Jock. *Raising Daisy Rothschild*. New York: Warner Books, 1979.

Letts, Elizabeth. *The Eighty-Dollar Champion*. New York: Ballantine Books, 2012. Tells the dramatic story of Snowman the horse and Harry de Leyer.

Lewis, Damien. *The Dog Who Could Fly*. New York: Atria Books, 2014. True story of a German shepherd, Antis, who was adopted by the Royal Air Force during World War II and joined in flight missions.

Lewis, Damien. *Judy: The Unforgettable Story of the Dog Who Went to War and Became a True Hero*. New York: Quercus, 2014. Incredible tale of Judy the dog a real survivor of World War II as a prisoner of war and other events.

Linden, Eugene. "Can Animals Think?" *Time,* August 29, 1999.

Linden, Eugene. *Silent Partners: The Legacy of the Ape Language Experiments*. New York: Ballantine Books, 1986.

Lortie, Bret. "Tails at Ground Zero (Bulletins)." *Bulletin of the Atomic Scientists*. March-April, 2002

Macejko, Christina. "Pets in the White House Have Affected U.S. History; Politics: Pups Prove Popular When One Faces a Scandal or Needs a Few Votes." *DVM Newsmagazine,* October 2008, p. 22+. *General OneFile,* http://link.galegroup.com/apps/doc/A188794029/GPS?u=vol_e7&sid=GPS&xid=7b8e29e8.

Marsh, Melissa Amateis. "Fido Goes to War." *America in WWII,* April 2015, p. 10+. *General OneFile,* http://link.galegroup.com/apps/doc/A436798843/GPS?u=vol_e7&sid=GPS&xid=7a7154ec.

Myron, Vicki, and Brett Witter. *Dewey: The Small-Town Cat Who Touched the World*. New York: Grand Central Publishing, 2008.

National Aeronautics and Space Administration. "A Brief History of Animals in Space."

Nicoletti, Jackie. "Top 10 Heroic Animal Stories in Australia" *Australian Geographic,* April 13, 2017.

Orleans, Susan. *Rin Tin Tin: The Life and the Legend*. New York: Simon & Schuster, 2011. Detailed and entertaining story of Rin Tin Tin.

Orr, Aileen. *Wojtek the Bear: Polish War Bear*. Edinburgh: Birlinn Ltd., 2014. The unbelievable tale of the Polish bear and his World War II adventures.

Orr, Andrew, and Angus Whitson. *Sea Dog Bamse: World War II Canine Hero*. Edinburgh: Birlinn, 2012.

Owen, Linda. "Our Presidents' Pets." *World and I,* December 2004.

Paietta, Ann, and Jean Kauppila. *Animals on Screen and Radio: An Annotated Sourcebook*. Metuchen, NJ: Scarecrow Press, 1994.

Passarello, Elena. *Animals Strike Curious Poses: Essays*. Louisville, KY: Sarabande Books, 2017. Essays about famous animals from history written in eccentric styles.

Perry, Roland. *Bill the Bastard*. Sydney, N.S.W.: Allen & Unwin, 2012.

Plumb, Christopher. *Georgian Menagerie: Exotic Animals in Eighteenth-Century London*, London: I.B. Tauris & Co., 2015. Exotic animals in London.

Redmond, Shirley Raye, and Lois Bradley. *Blind Tom: The Horse Who Helped Build the Great Railroad*. Missoula, MT: Mountain Press, 2009.

Ridley, Glynis. *Clara's Grand Tour: Travels with a Rhinoceros in Eighteenth-Century Europe*. New York: Grove Press, 2004. True tale of Clara and her travels through Europe.

Robinson, James. *Larry Diaries: Downing Street—the First 100 Days*. New York: Simon & Schuster, 2011.

Rossouw, Sarel and Wikipedia. "Jackie, The Baboon Mascot of 3 SAI During The Great War, 1914–1918." *Military History Journal,* Vol 16, December 2, 2013.

Schmidle, Nicholas. "Getting Bin Laden." *The New Yorker,* August 8, 2011.

Seguin, Marilyn. *Dogs of War: And Stories of Other Beasts of Battle in The Civil War.* Boston: Branden Publishing Co,. 1998. Civil War tales of friendships of animals between the soldiers.

Seton, Ernest Thompson. *Famous Animal Stories, Animal Myths, Fables, Fairy Tales, Stories of Real Animals.* New York: Brentano's, 1932. Selected extracts from books about animals.

Sher, Lynn. *Tall Blondes, A Book about Giraffes.* Kansas City: Andrews McMeel, 1997.

Shreeve, James. "Oliver's Travels." *The Atlantic Monthly,* October 2003.

Sime, Harriet. "Did This War Hero Donkey Make an Ass of the RSPCA?" *Daily Mail* (London, England), December 20, 2014.

Simons, John. *Rossetti's Wombat.* London: Middlesex University Press, 2008.

Sisson, Terence. *Just Nuisance AB, His Full Story.* Cape Town: W. J. Flesch & Partners Ltd., 1986.

Smith, Steven Trent. "A Few Good Marines: Dogs in Wartime." *World War II Magazine.* February 16, 2017.

Smith, Steven Trent. "A Few Good Marines: In the Pacific, American Warriors Learned How to Let Slip the Dogs of War." *World War II,* March-April 2014, p. 26+. *General OneFile,* http://link.galegroup.com/apps/doc/A356038357/GPS?u=vol_e7&sid=GPS&xid=89ae8459.

Steinbeck, John. *Travels with Charley: In Search of America.* New York: Penguin Books, 1962.

Stoddard, Maynard Good. "Presidential Pets." *Saturday Evening Post,* July-August 1992.

Street, Peter. *Animals in the First World War.* Stroud, Gloustershire, UK: Pitkin, 2016.

Stroud, Patricia. *Caesar: The Anzac Dog.* Auckland (N.Z.): HarperCollins Publishers, 2009.

Thomas, Lillian. "Civil War's Dog, Jack Saluted After 7 Score, 4 Years." *Pittsburgh Post-Gazette,* August 10, 2005.

Tiger, Caroline. *General Howe's Dog: George Washington, the Battle for Germantown and the Dog Who Crossed Enemy Lines.* New York: Chamberlain Bros., 2005. Entertaining true tale of how a dog played a role during the Revolutionary War.

Trumble, Angus. "Rossetti, Morris and the Wombat." *Art and Australia,* Autumn 2012.

Trumble, Angus. "Rossetti's Wombat (An Arena Christmas Cracker)." *Arena Magazine,* December 2002.

Van Emden, Richard. *Tommy's Ark: Soldiers, Their Animals and the Natural World in the Great War.* London: Bloomsbury, 2010.

"A Wag in the Tale." *The Australian Magazine* (New South Wales, Australia), October 19, 2013, p. 22. *Academic OneFile,* http://link.galegroup.com/apps/doc/A345949938/GPS?u=vol_e7&sid=GPS&xid=ebe94e10.

Walker, Sam. "War Hero Who Could Raise a Hoof in Salute," *Daily Mail* (London, England), March 24, 2012.

Walker-Meikle, Kathleen. *The Dog Book: Dogs of Historical Distinction.* Oxford: Old House Books, 2014. Short tales of dogs throughout history.

Walker-Meikle, Kathleen. *Medieval Pets.* Rochester: Boydell & Brewer, 2012.

Walters, Jennifer Marino. "War Dogs: Courageous Canines Use Their Noses to Help Soldiers." *SuperScience,* October 2011, p. 14+. *General OneFile,* http://link.galegroup.com/apps/doc/A269337526/GPS?u=vol_e7&sid=GPS&xid=744102ac.

Wasson, Donald L. "Bucephalus." *Ancient History Encyclopedia,* October 6, 2011.

Watson, Bruce. "The Dogs of War: There's a Special Quality in Some Dogs—Call It Loyalty, Heroism or Just Plain Courage—That Comes Alive Under Fire." *Smithsonian,* December 2000, p. 100. *General OneFile,* http://link.galegroup.com/apps/doc/A69861729/GPS?u=vol_e7&sid=GPS&xid=4602d870.

Weatherwax, Rudd B., and John H. Rothwell. *The Story of Lassie: His Discovery and Training from Puppyhood to Stardom.* New York: Duell, Sloan and Pearce, 1950.

Weintraub, Robert. *No Better Friend: One Man, One Dog, and Their Extraordinary Story of Courage and Survival in World War II.* New York: Little, Brown and Co., 2015. The tale of Judy and her story of friendship and courage.

Wellerstein, Alex. "Remembering Laika, Space Dog and Soviet Hero." *The New Yorker,* November 3, 2017.

Whitaker, Julie. *The Horse: A Miscellany of Equine Knowledge.* Lewes East Sussex, UK: Ivy Press, 2007.

Whitson, Angus. *Sea Dog Bamse: World War II Canine Hero.* Edinburgh: Birlinn, 2012. True story of Bamse.

Whitt, Kelly Kizer. "Reluctant Astronauts: How Other Creatures Paved the Way for Human Space Travelers." *Astronomy,* April 2001, p. 42. *General OneFile,* http://link.galegroup.com/apps/doc/A75106292/GPS?u=vol_e7&sid=GPS&xid=b3d02f18. Ac.

Williams, James Howard. *Bandoola: The Story of an Elephant.* New York: Penguin Books, 1962.

Woodward, Daisy. "Salvador Dalí's Ocelot." *AnOther,* January 23, 2013.

"World Remembers USSR's Laika, First Dog in Outer Space." *Information Company,* November 4, 2017. *General OneFile,* http://link.galegroup.com/apps/doc/A513216147/GPS?u=vol_e7&sid=GPS&xid=3af329ec.

Worman, Charles. *Civil War Animal Heroes: Mascots, Pets and War Horses.* Lynchburg, VA: Schroeder Publications, 2011. Stories of animals during the Civil War.

Wynne, William A. *Yorkie Doodle Dandy: A Memoir.* Denver: Top Dog Enterprises, 1996.

Zucchero, Michael. *Loyal Hearts: Histories of American Civil War Canines.* Lynchburg, VA: Schroeder Publication, 2011. Tales of dogs during the Civil War.

Index